Sensor and Data Fusion for Intelligent Transportation Systems

Sensor and Data Fusion for Intelligent Transportation Systems

Lawrence A. Klein

SPIE PRESS
Bellingham, Washington USA

Library of Congress Cataloging-in-Publication Data

Names: Klein, Lawrence A., author.
Title: Sensor and data fusion for intelligent transportation systems / Lawrence A. Klein.
Description: Bellingham, Washington, USA : SPIE Press, [2019] | Includes bibliographical references.
Identifiers: LCCN 2019001144 | ISBN 9781510627642 (softcover) | ISBN 1510627642 (softcover) | ISBN 9781510627659 (PDF) | ISBN 1510627650 (PDF) | ISBN 9781510627666 (ePub) | ISBN 1510627669 (ePub) | ISBN 9781510627673 (Kindle) | ISBN 1510627677 (Kindle)
Subjects: LCSH: Intelligent transportation systems. | Multisensor data fusion. | Motor vehicles–Automatic control. | Traffic congestion. | Algorithms.
Classification: LCC TE228.3 .K537 2019 | DDC 388.3/12–dc23
LC record available at https://lccn.loc.gov/2019001144

Published by

SPIE
P.O. Box 10
Bellingham, Washington 98227-0010 USA
Phone: +1 360.676.3290
Fax: +1 360.647.1445
Email: books@spie.org
Web: http://spie.org

Printed in the United States of America.
First Printing.
For updates to this book, visit http://spie.org and type "PM305" in the search field.

SPIE.

To Jonathan, Amy, Gregory, and the Magnificent Seven:
Maya, Theo, Coco, Casper, Cassie, Tessa, and Johnny

Table of Contents

Preface

Sensor and Data Fusion for Intelligent Transportation Systems was prepared to give undergraduate and graduate students, researchers, and traffic management professionals a multidisciplinary description of sensor and data fusion and the benefits it brings to the transportation community, especially for intelligent transportation systems (ITS). Although sensor and data fusion processes are introduced through the lens of the military-oriented U.S. Joint Directors of Laboratories (JDL) data fusion model and the Data Fusion Information Group (DFIG) enhancements, analogous language for its application to traffic management is afforded at each data fusion processing level. This model was selected because of its ability to expose all of the generally acknowledged facets of data fusion.

The book focuses on data fusion functions and data processing algorithms. Sensor fusion architectures, although discussed, are not treated in as much detail. The discussions explore techniques that enhance the interpretation of information gathered from a diverse mixture of sensors and other data sources (for instance, floating cars, connected and cooperative vehicles, self-driving vehicles, cellular telephone messages, mobile device locations accessible through Bluetooth® communications, location data from global navigation satellite system devices, and automatic license plate and toll-tag readers) that help characterize the traffic environment. Its often-complex nature is exacerbated by a mix of different types of vehicles, changes in traffic flow characteristics, appearance of unexpected objects such as pedestrians darting across a roadway, inclement weather, vehicles changing lanes, and roadside structures or poor visibility conditions that interfere with the normal observation of traffic patterns and the gathering of needed data. Bayesian inference, Dempster–Shafer evidential reasoning, and Kalman filtering data fusion algorithms that combine data from infrastructure-based sensors and other sources are described in detail with illustrative examples of their application to ITS. Other algorithms applicable to fusion of traffic data are also discussed, including parametric templates, artificial neural networks, cluster algorithms, voting methods, knowledge-based expert systems, and fuzzy logic.

The manuscript is based, in part, on short courses and semester-length courses taught by the author at several universities and conferences for many years. The impetus for the book came from a desire to prepare a treatment of selected subjects pertaining to sensor and data fusion for novice and more experienced practitioners from universities; transportation institutes; and local, regional, state, and multi-state or multi-national agencies and consulting companies. The intent is to provide an understanding of the functions of multilevel data fusion processes suitable for ITS, knowledge of the kinds of information and parameters required to implement particular data fusion algorithms, and recognition of pertinent issues that may affect the operation of a data fusion system. Personnel engaged in research may be particularly interested in the sections that deal with ongoing needs in data fusion research.

Readers desiring additional information concerning sensor and data fusion are referred to *Sensor and Data Fusion: A Tool for Information Assessment and Decision Making*, Second Edition, SPIE Press, Bellingham, Washington (2012) [doi: 10.1117/3.928035], which is written by the author. Further material regarding freeway traffic management centers, ITS, sensors and other data acquisition devices, data requirements, sensor testing, automated vehicles and vehicle systems, connected vehicle programs in the U.S. and elsewhere, systems engineering, and National Intelligent Transportation System Architectures is found in *ITS Sensors and Architectures for Traffic Management and Connected Vehicles*, Taylor and Francis, Boca Raton (2018), which is also written by the author.

I wish to thank Maijian Qian for assisting in verifying several of the equations that appear in Chapter 5; Dara Burrows, Senior Editor SPIE Press, and Tim Lamkins, SPIE Press Manager, who facilitated the editing and publication of the book and afforded me their usual courtesy; and finally, the reviewers for providing valuable feedback that greatly enhanced the material that appears here.

Lawrence A. Klein
June 2019

Acronyms

AC	alternating current
ADAS	advanced driver assistance systems
AID	automatic incident detection
AIS	automatic identification systems
ALPR	automatic license-plate reader
ANN	artificial neural network
ATIS	advanced traveler information systems
ATMS	advanced transportation management systems
AVI	automatic vehicle identification
BOA	bisector of area
C-ITS	Cooperative Intelligent Transport Systems (European Union)
CA	certificate authority
CA	constant acceleration
CART	classification and regression tree
CAS	collision avoidance systems
CB	citizen band
CCP	critical collision probability
CCR	correct classification rate
CJ	constant jerk
CL	change lane
CL	constant location
COM	center of mass
COTS	commercial-off-the-shelf
CRL	certificate revocation list
CSAWM-JDL	combined situational awareness model and JDL model
CV	constant velocity
CVO	commercial vehicle operation
DAG	directed acyclic graph
DBMS	database management systems
DBSCAN	density-based spatial clustering of applications with noise
DFIG	Data Fusion Information Group
DRGS	dynamic route guidance systems

DSRC	dedicated short-range communications
DSS	decision support system
DT	decision tree
EKF	extended Kalman filter
EU	European Union
FASUR	Fuse-Act Situational User Refinement
FCD	floating car data
FHWA	Federal Highway Administration
FOM	first of maxima
FRESIM	freeway simulation
GPR	ground-penetrating radar
GPS	global positioning system
HCI	human–computer interface
HIT	hard-information timeliness
HOG	histogram of oriented gradients
HM	height maximum
IBR	information benefit ratio
IC	information confidence
IEEE	Institute of Electrical and Electronics Engineers
IFE	information fusion effectiveness
IG	information gain
ILD	inductive loop detector
IMM	interacting multiple model
INS	inertial navigation system
IQ	information quality
IR	infrared
IT	information timeliness
ITS	intelligent transportation systems
JDL	Joint Directors of Laboratories
KL	keep lane
KNN	k^{th} nearest neighbor
LOM	largest of maxima
LoM	last of maxima
MaaS	Mobility as a Service
MAC	media access control
MAP	maximum *a posteriori*
MCS	multiple-classifier system
MD	metal detector
MEBN	multi-entity Bayesian network
MIM	Midtown in Motion
MLFF	multilayer feedforward
MMI	man–machine interface
MMSE	minimum mean-square error

MoE	measure of effectiveness
MOM	mean of maxima
MoP	measure of performance
NATO	North Atlantic Treaty Organization
NB	naïve Bayes
NHTSA	National Highway Traffic Safety Administration
NMR	noise-to-maneuver ratio
NN	nearest neighbor
NRR	noise reduction ratio
OD	origin–destination
OIT	overall information timeliness
pdf	probability density function
PFS	parametric free space
PKI	public key infrastructure
pmf	probability mass function
PNN	probabilistic neural network
PRM	perceptional reasoning model
QoS	quality of service
QZSS	Quasi-Zenith Satellite System (Japan)
RAL	risk awareness level
RMF	risk management framework
ROI	reliability of information
ROS	reliability of source
SAW	situation awareness
SAWM	situational awareness model
SCATS	Sydney Coordinated Adaptive Traffic System
SCM	security and credentials management
SIT	soft-information timeliness
SNR	signal-to-noise ratio
SOC	sum of center
SOM	smallest of maximum
STANAG	standardization agreement
SVM	support vector machine
SVR	support vector regression
TOC	traffic operations center
TRR	transient response ratio
TSM&O	Transportation Systems Management and Operations
TTC	time-to-collision
TTCCP	time-to-critical-collision probability
UFM	user fusion model
UKF	unscented Kalman filter
V2I	vehicle-to-infrastructure
V2P	vehicle-to-pedestrian

V2V	vehicle-to-vehicle
V2X	vehicle-to-everything
VANET	vehicular *ad hoc* network
WAAS	wide-area augmentation system
WIM	weigh-in-motion

Chapter 1
Introduction

Transportation systems require reliable and accurate data for monitoring and managing operations that maximize the safety and efficiency of the roadway network. The proliferation of connected (or cooperative) and automated vehicles, which are capable of providing data regarding vehicle speeds and volumes, congestion and collision avoidance maneuvers, and inclement weather and other potentially dangerous road conditions, create both a major opportunity and a major challenge for public agencies and private companies that support intelligent transportation systems (ITS). The opportunity arises from the diversity of data available to traffic and transportation management personnel, while the challenge arises from the need to gather, process, analyze, and store the vast quantities of data that will flow to the traffic management agencies, especially as big data become available. These emerging and fertile data sources, along with the spread of Bluetooth® and IP-based (cellular and Wi-Fi) communications technologies, will increase the travelers' proclivity for accurate road traffic information.

Predictions by Gettman, Hales, Voss, and Tumati[1] for the connected city of 2021 forecast growth of 400% in connected travelers, 300% in connected vehicles, and 25% in connected infrastructure. As the connected city continues to grow, the amount of data stored within the system will need to grow from terabytes to petabytes in 10 years—explaining the importance of scalability (i.e., a system's ability to increase and/or decrease storage resources as needed). If all of this emerging data (such as from connected travelers, connected vehicles, and new types of connected infrastructure) related to traffic operations are stored, the cumulative storage of a typical traffic management agency could be in the many thousands of terabytes by 2026.[2] A challenge is to develop data fusion methods that can cope with these vast quantities of data.

Sensors that monitor traffic flow at a given point are often ineffective in supplying the data required by modern transportation management systems. For example, origin–destination (OD) pair data needed for planning purposes and vehicle density studies are not readily available from point sensors.

Global positioning system (GPS) and other global navigation satellite systems' location devices, cell phone tracking through media access control (MAC) address readers, probe vehicles, automatic license-plate readers (ALPRs), toll-tag readers, and trucking industry transponders are increasingly supplementing the information provided by conventional traffic flow sensors.

When multisource data are complementary in nature and an output from one data source does not imply nor is implied by an output from another (i.e., the data sources are conditionally independent), data fusion can provide operational benefits to traffic management personnel. These benefits include utilizing the data to obtain a more accurate description of the road and traffic conditions by decreasing the uncertainty present in the individual sources of data, extending spatial and temporal coverage areas, accessing data not usually available from conventional roadside sensors such as OD pairs and connected vehicle data (for instance, weather and road conditions from in-vehicle sensors that automatically turn on windscreen wipers or initiate traction control measures), and providing the ability to function in inclement weather. The fusion of data and information from multiple sources is consequently a well-adapted answer to the operational needs of traffic management centers and traffic information providers, allowing them to achieve their goals more effectively.[3] Data fusion opportunities also exist for processing archived data such as traffic volumes by time-of-day, day-of-week, month, or season, and recurring special events. This offline information, together with real-time sensor data, is often useful in predicting traffic trends and forecasting the need for new roadways or other travel modalities.

Sensor and Data Fusion for Intelligent Transportation Systems introduces readers to the roles of several data fusion processes as defined by the Joint Directors of Laboratories (JDL) data fusion model and the Data Fusion Information Group (DFIG) enhancements, data fusion algorithms, and noteworthy applications of data fusion to ITS. Additionally, the monograph offers detailed descriptions of three of the widely applied data fusion techniques and their relevance to ITS (namely, Bayesian inference, Dempster–Shafer evidential reasoning, and Kalman filtering), and indicates directions for future research in the area of data fusion. The focus is on data fusion algorithms rather than on sensor and data fusion architectures, although the book does summarize factors that influence the selection of a fusion architecture and several architecture frameworks.

1.1 Applications to ITS

Effective design of an ITS requires a systems approach that incorporates sensors and other data-gathering devices and a variety of communications technologies into the concept exploration, architecture selection, hardware and data processing design, testing, operation, and performance evaluation of the fabricated system. Systems engineering helps assure that a proposed architecture will satisfy the goals and objectives of the project. Typical projects that benefit from this method include freeway designs that employ ramp metering, information dissemination to travelers, managed lanes, and active traffic management; arterials that utilize traffic adaptive signal control, signal priority for transit, freight, and emergency vehicles, and parking guidance; integrated corridors that coordinate and seek to optimize traffic flow on arterials and limited-access highways; road-weather systems; and the construction and staffing of transportation management centers, especially those that collocate personnel from different agencies.

Several data fusion algorithms are already prevalent in ITS applications. These include Bayesian inference, Dempster–Shafer evidential theory and some of its modifications, artificial neural networks, fuzzy logic, knowledge-based expert systems, and vehicle and pedestrian tracking based on the Kalman filter or extended Kalman filter (EKF), Monte Carlo techniques, and particle filters. The Bayesian and Dempster–Shafer approaches belong to the class of feature-based parametric algorithms. They directly map parametric data (e.g., features) into a declaration of identity. Physical models are not used. Artificial neural networks belong to the class of feature-based information theoretic techniques that transform or map parametric data into an identity declaration. No attempt is made to directly model the stochastic aspects of the observables. Fuzzy logic and knowledge-based expert systems are examples of cognitive-based approaches that attempt to emulate and automate the decision-making processes used by human analysts. The Kalman filter and its nonlinear-motion counterparts are examples of physical models since the kinematics of the objects being tracked are modeled. Physical models replicate object discriminators—in this case, position, velocity, and sometimes acceleration—that are easily observable or calculable.

ITS subsystems include advanced transportation management systems (ATMS), automatic incident detection (AID) (a subset of ATMS), advanced traveler information systems (ATIS), advanced driver assistance systems (ADAS), and commercial vehicle operations (CVO), all of which gather data and information from different sources. Data fusion techniques can therefore be developed to combine the data and obtain a better understanding of what the data represent, e.g., the types of objects or the situation (for instance, a roadway incident) giving rise to the data.

Incident detection algorithms that automatically detect incidents, accidents, and other road events requiring emergency responses have existed for more than three decades. Most of the algorithms rely on loop detector data. However, these algorithms exhibit mixed success. Interest in incident detection algorithms has renewed partly because of the availability of new sensors and data sources such as probe and connected vehicles and cellular telephone reporting. However, data fusion involving combinations of hard (i.e., sensor) and soft (i.e., voice) data is not without risks and is a subject of current research. Notwithstanding the preceding comment, AID is typical of the class of problems that can be enhanced by data fusion techniques.

ATIS employ a variety of automatic data collection techniques to assist in understanding traffic conditions and derive relevant indicators that support traveler guidance. Traveler information is often presented as travel time or, in the worst case, as a road closure or extended delay due to an incident. In this context, travel time is used as a measure of impedance (or cost) for route choice strategies. However, conventional traffic sensors that measure the prevailing traffic conditions on an urban road may be ineffective at this task because of a sparsity of sensors or their inoperability. The proliferation of alternative data and information sources (e.g., surveillance cameras, GPS, MAC address readers, ALPRs, connected and automated vehicles, and cell phone reports) provide data that supplement traditional sensor measurements. These complementary devices also have the potential to improve the accuracy of travel-time estimates. As a result, travel-time estimation becomes a candidate data fusion application.

Improving traveler safety is a primary function of ITS. The increased availability of ADAS and collision avoidance systems (CAS) is indicative of the growth in active safety devices that complement traditional passive ones such as seat belts and air bags. ADAS help alert a driver to traffic and other hazards surrounding the vehicle in pre-crash situations. Their input data frequently come from a variety of sensors including radar, lidar, ultrasonic, and video imaging sensors. Fusion is exploited to combine these data and alert the driver to potentially dangerous situations. CAS have the ability to notify not only drivers of potential incident situations, but also traffic management personnel who can take actions such as reducing speed limits on the affected highways or lanes, closing lanes, and notifying state highway police to implement a traffic break or initiate other accident-preventive procedures.

An important item in the field of transportation planning and control is OD estimation from counts undertaken on specific links of the transportation network. An estimation of a most likely OD matrix is then derived from the counts. Traffic flow forecasting is of increasing importance to traffic surveillance, facility management, and departments that plan for new roadways. Many traffic flow prediction schemes of the past were based on classic autoregressive models, especially time series techniques. Connected vehicle, GPS, MAC address, and ALPR data are now available to enhance the acquisition of OD pair information.

Modern transportation systems require accurate information concerning the position and orientation of vehicles to forecast congestion and incident precursors such as wrong-way drivers. This application has been abetted by the ubiquitous nature of GPS for location and navigation services. However, when satellite signals are blocked by tall buildings and trees, or are corrupted by electromagnetic interference or refraction as they propagate through the atmosphere, loss or degradation of the GPS signal occurs. In such situations, the estimation of position is degraded at best or, at worst, is impossible to obtain as the device is unable to acquire a signal. Inertial navigation systems (INS) that determine the location of a vehicle rely on dead reckoning and can be used to complement GPS data when the latter are degraded or unavailable. However, INS are subject to integration drift caused by the accumulation of small errors in the measurement of acceleration and angular velocity that manifest as larger errors in the position estimate. Data fusion offers a complementary approach that combines the benefits of the GPS and INS techniques, namely, calibration of the INS by the GPS signals when they are available, and position and angle updates provided by the INS when the GPS signal is blocked.

Another area closely related to ITS is autonomous or self-driving vehicles. The need for data fusion in these vehicles was noted by Raj Rajkumar, Professor of Electrical and Computer Engineering at Carnegie Mellon University, at a forum sponsored by the National Academy of Engineering in 2018.[4] Prof. Rajkumar remarked that some companies are not relying on GPS to provide location data to the autonomous vehicles they manufacture because there are many GPS-denied environments, such as under bridges, inside tunnels, in urban canyons, and in densely forested areas. Instead, these companies are working toward the identification of landmarks to locate the vehicle. However, certain places have very few landmarks. Rajkumar observed that "There are pros and cons to every approach you can take. My own philosophy is that you need to fuse together as many things as you can [to] get the best of all possible worlds."[4] Furthermore, the components of an autonomous vehicle can fail, so redundancy must be built into the vehicle. This presents another opportunity to incorporate data fusion into such a vehicle.

1.2 Data, Information, and Knowledge

Data, information, and knowledge are terms found throughout this book. Their relation to each other is illustrated in Fig. 1.1 in the form of a triangle whose foundation is the data that evolves into information and finally into knowledge through further processing, interpretation, and comprehension.[5] Data are the individual observations, measurements, and primitive messages from the lowest level of abstraction. Data are obtained from human communication, text

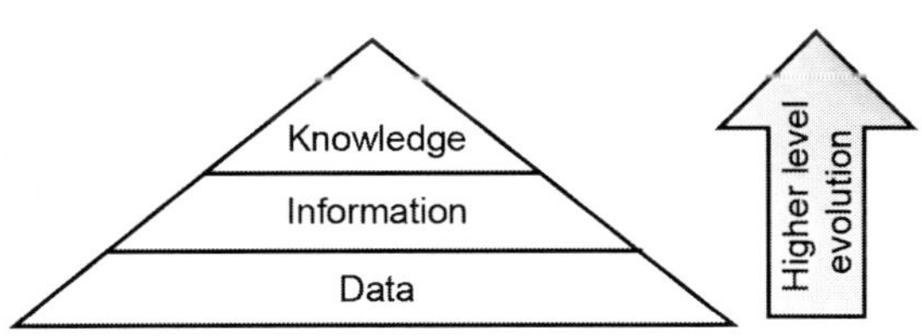

Figure 1.1 Evolution of data to information and knowledge.

messages, electronic queries, or devices that sense phenomena. Evidence consists of relevant data or specific elements of the overall data set.

Information is represented by organized sets of data. Organization may occur through sorting, classifying, and indexing and linking data to place data elements in relational context for subsequent searching and analysis.

Finally, knowledge, or foreknowledge (i.e., predictions or forecasts), evolves from information that is analyzed, understood, and explained. Once understood, knowledge provides a degree of comprehension of the static and dynamic relationships among data objects, the ability to model structures, and an understanding of past and future behavior of those objects.

1.3 Summary of Book Contents

This book contains five additional chapters. Chapter 2 includes several common definitions of sensor and data fusion, and presents an argument for its application to traffic management. The relevance of multisensor data fusion is due to the following factors: its value in combining and interpreting information gathered from a complex environment characterized by the presence of different types of vehicles; the often rapid changes in traffic flow characteristics; unexpected objects such as debris, pedestrians, or animals darting across a roadway; vehicles changing lanes; and roadside structures or weather effects that interfere with the normal observation of traffic patterns and the gathering of needed data. Chapter 2 describes the processes typically associated with data fusion and their benefits with the aid of the JDL data fusion model and the DFIG augmentations. Although this model was originally developed with military-oriented applications in mind, its concepts and approaches are shown to be directly applicable to traffic management.

The traffic management data fusion strategies discussed in the chapter utilize data from traffic flow sensors installed along the roadway; devices that access toll-tag and electronic screening and truck transponders, ALPRs, and Bluetooth-device MAC address readers used to track and re-identify vehicles; floating cars that provide a traffic management center emissions information in addition to normal traffic flow parameters linked to the vehicle's location via GPS; crowdsourcing applications and personal device monitoring that measure travel behavior of pedestrians and bicyclists; and, in the not too distant future, connected and cooperative vehicle data associated with vehicle-to-vehicle (V2V), vehicle-to-infrastructure (V2I), and vehicle-to-pedestrian (V2P) applications. A summary of factors that influence the selection of a sensor and data fusion architecture and several architecture classification schemes are provided, although this is not the primary focus of the chapter. The emphasis is on data fusion algorithms and processes, and their applications to ITS and traffic management. Consequently, the discussions of the taxonomies and several of the algorithms used for object detection, classification, identification, and state estimation and tracking utilize transportation-related examples to illustrate their relevance. The higher-level data fusion processes of situation and impact assessment, performance refinement, and user refinement identify behavior patterns (in this case, traffic behavior), associate entities and events, predict future behaviors and their time frames, assess the situation generating the collected data, refine the fusion process, and address issues concerning human interpretation of the results of the data fusion process. The chapter concludes by describing a variety of sensor and data fusion applications to ITS and its subsystems.

Chapter 3 discusses the underlying principles of Bayesian inference, the assumptions typically associated with its application, and several examples that illustrate its relevance to traffic management. Common prerequisites for Bayesian analyses are the ability to: (1) compute or model the likelihood functions for each sensor or information source and object in the scenario of interest, (2) use a set of mutually exclusive hypotheses and calculate the *a priori* probabilities that the hypotheses are true, (3) discard past evidence or data once the posterior probability is calculated for the current period of interest, and (4) obtain data from conditionally independent sensors. Bayesian inference is widely employed for detection, classification, and identification of objects and events related to traffic management. It provides travel-time estimation, automatic incident detection, and decision support. The examples described in the chapter are identification of vehicle type with emission spectra data collected by a sensor, incident detection utilizing an influence diagram and a joint sensor report, truck classification using a multiple classifier system in which two of the processes are Bayesian, and maneuver-based trajectory prediction and criticality assessment by means of causal Bayesian networks.

Chapter 4 examines Dempster–Shafer evidential theory, a pseudo-probability–based data fusion classification algorithm. This method finds application when the sensors (or, more generally, the information sources) contributing data cannot associate a 100% probability of certainty to their output decisions. The sensors must function as independent sources of information concerning the presence of the objects or events of interest. Knowledge from multiple sensors about events (called propositions) is combined using Dempster's rule to find the intersection or conjunction of the propositions and their associated probability, which is expressed as probability mass, sometimes called a basic probability assignment. The chapter summarizes several modifications to the original Dempster–Shafer theory that have been proposed to accommodate situations where the information sources are in conflict. Perhaps the most difficult part of applying Dempster–Shafer theory in its original or modified forms is obtaining probability mass functions. Five methods for developing these functions are discussed. An example is given of the application of Dempster–Shafer reasoning to create a decision support system that enhances a traffic manager's understanding of the conditions that give rise to the collected data.

The Kalman filter described in Chapter 5 provides an optimal state estimate for linear systems as long as the dynamics of the tracked object and measurement noise are accurately modeled. The filter estimates an object's state (i.e., its position, velocity, and acceleration) at some future time, e.g., the predicted time of the next observation, and then updates that estimate using noisy measurements of the state. Process noise present in the kinematics of the object must also be modeled and of sufficient magnitude to keep the Kalman gain large enough to ensure that the tracker does not ignore the current measurement data and simply dead reckon the track based on past history. Moreover, the Kalman filter offers an estimate of the tracking error statistics through the state error covariance matrix. Also discussed are the α-β filter, which is applicable when the tracked object moves with constant velocity, an interacting multiple model to describe potential vehicle motion on a roadway, and the extended Kalman filter, which is applicable to nonlinear systems.

Chapter 6 reviews the state of the practice of data fusion in traffic management applications and the need for continued research in several areas that include reliability and quality of input data to the fusion system, security of data fusion systems, fusion of sensor or hard data with human-generated or soft data, fusion system performance assessment, the adaptive nature of measures of performance (MoPs), dependency of MoPs on the data fusion scenario, obtaining accurate ground truth data, use of commercial database management and operating systems tools, and designing for worst-case scenarios to properly specify processing and communications requirements. The chapter concludes with a summary of the information needed to apply Bayesian inference, Dempster–Shafer evidential theory, artificial neural networks, voting logic, fuzzy logic, and Kalman filtering data fusion algorithms to object detection, classification, identification, and state estimation.

References

1. D. Gettman, K. Hales, A. Voss, A. Toppen, and B. Tumati, "Integrating Emerging Data Sources into Operational Practice: State of the Practice Review," FHWA-JPO-16-424, Kimley Horn and Associates, Phoenix, Dec. (2016).
2. D. Gettman, A. Toppen, K. Hales, A. Voss, S. Engel, and D. El Azhari, "Integrating Emerging Data Sources into Operational Practice—Opportunities for Integration of Emerging Data for Traffic Management and TMCs," FHWAJPO-18-625, Kimley Horn and Associates, Phoenix, Nov (2017).
3. N.-E. El Faouzi and L. A. Klein, "Data Fusion in Intelligent Traffic and Transportation Engineering: Recent Advances and Challenges," in

Multisensor Data Fusion: From Algorithm and Architecture Design to Applications, H. Fourati, Ed., Chapter 32, CRC Press, Boca Raton, Florida (2015).

4. National Academy of Engineering, *Autonomy on Land and Sea and in the Air and Space: Proceedings of a Forum*, The National Academies Press, Washington, D.C., pp. 2–3 (2018) [doi: 10.17226/25168. Also available at http://nap.edu/25168. Accessed Aug. 7, 2018].
5. J. Roy, "Combining elements of information fusion and knowledge-based systems to support situation analysis," *Proc. SPIE* **6242**, 624202 (2006) [doi: 10.1117/12.663896].

Chapter 2
Sensor and Data Fusion in Traffic Management

Multisensor data fusion offers many benefits to traffic management. It aids in the interpretation of data gathered from a complex environment characterized by the presence of different types of vehicles, unexpected objects such as pedestrians and animals darting across a roadway, vehicles changing lanes, challenging road geometries, inclement weather, and roadside structures that interfere with the normal observation of traffic patterns and the gathering of needed data.[1–3] Data fusion processes and algorithms that combine data from infrastructure-based sensors and other sources offer the potential for overcoming these impediments and providing correct interpretations of the traffic conditions at hand. Furthermore, sensor and data fusion can assist traffic management personnel in gaining insight into what caused the observed events and their impacts on the occurrence of future events that could affect traffic flow.[1,3,4]

This chapter concentrates on two areas: (1) sensor and data fusion in general and (2) specific applications of data fusion to ITS and traffic management. The first sections describe several commonly encountered definitions of sensor and data fusion, ways in which sensor and data fusion enhance the effectiveness of traffic management strategies, the ever-expanding sources available for traffic data, and the selection and classification of fusion architectures. The U.S. Department of Defense JDL data fusion model and DFIG enhancements are utilized to describe the functions and inferences that are typical of data fusion processes.[5–9] A taxonomy is offered for object detection, classification, and identification algorithms, followed by the taxonomy for the state estimation algorithms utilized to track objects. The algorithms and procedures identified in these taxonomies (referred to as Level 1 processing in the JDL fusion model) are then briefly discussed along with applications to traffic management. Klein,[2] Waltz and Llinas,[5] Khaleghi et al.,[10] and Castanedo[11] provide additional information concerning these techniques. The choice of a data fusion algorithm is often made by considering which method makes correct inferences concerning the situation of interest and the availability of the required computer resources and algorithm input parameters. Several sections are devoted to describing the functions of JDL data fusion processing Levels 2–5 and their relevance to traffic management. The final section presents examples of sensor and data fusion applications to ITS.

2.1 What is Meant by Sensor and Data Fusion?

According to Waltz and Llinas, "Data fusion is a process performed on multisource data at several levels, each of which represents a different level of abstraction of data."[5] Data fusion is concerned with the following:

1. Representation of information within a computational database, particularly the information gained through data fusion, and
2. Presentation of this information in a manner that supports the required decision processes when a human operator or decision maker is involved.

Data fusion should not be the goal or end result of a transportation management strategy. Rather the goal is to provide to a control system, in the form of a machine or a human, the information necessary to support automated or semi-automated decision making, such as in ITS applications where vehicle systems, drivers, or traffic management personnel may have to take corrective actions to ensure traveler safety and the smooth flow of traffic.

A variety of definitions of sensor and data fusion are found in the literature. The JDL model, perhaps the most widely cited, defines data fusion as "a multilevel, multifaceted process dealing with the automatic detection, association, correlation, estimation, and combination of data and information from single and multiple sources to achieve refined position and identity estimates, and complete and timely assessments of situations and threats and their significance."[5,12] The Institute of Electrical and Electronics Engineers (IEEE) Geoscience and Remote Sensing Society's definition is "the process of combining spatially and temporally indexed data provided by different instruments and sources in order to improve the processing and interpretation of these data." Boström, Andler, Brohede, et al. of the University of Skövde provide a definition in terms of information fusion as "the study of efficient methods for automatically or semi-automatically transforming information from different sensors and different points in time into a representation that provides effective support for human or automated decision making."[13] Dasarathy offers another definition of information fusion as encompassing "the theory, techniques, and tools conceived and employed for exploiting the synergy in the information acquired from multiple sources ... such that the resulting decision or action is in some sense better (qualitatively or quantitatively, in terms of accuracy, robustness, etc.) than would be possible, if these sources were used individually without such synergy exploitation."[14] These definitions offer different insights into the role of sensor and data fusion and reflect their diverse applications.

The terms *data fusion* and *sensor fusion* are often used interchangeably. Strictly speaking, data fusion is defined in the context of the preceding paragraph. Sensor fusion, then, refers to the use of more than one sensor or data source in a configuration that enables more accurate or additional data to be gathered about events or objects in the observation space of the sensors and the distribution of processing and communications resources among several fusion nodes or locations. For a number of reasons, multiple sensors may be needed to completely and continually monitor the observation space. For instance, some objects may be detected by one sensor but not another because each sensor may respond to a different signature-generation phenomenology or be affected by inclement weather in different ways. The signature of an object may be masked or otherwise hidden with respect to one sensor but not another; or one sensor may be blocked from viewing objects because of the geometric relation of the sensor to the objects in the observation space, but another sensor located elsewhere in space may have an unimpeded view of the object. In this case, the data or tracks from the sensor with the unimpeded view may be combined with past information (i.e., data or tracks) from the other sensor to update the state estimate of the object.[2]

2.2 Sensor and Data Fusion Benefits to Traffic Management

Figure 2.1 illustrates the benefits that multisensor fusion can bring to traffic management. Multisensor fusion assists in the interpretation of information gathered from a multifaceted and often complex environment that contains a

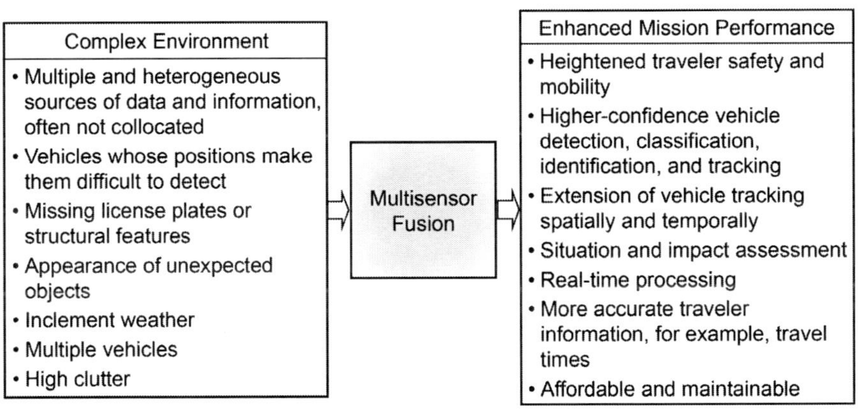

Figure 2.1 Enhancing mission performance in a complex environment with multisensor fusion.

variety of data sources, vehicles, and other objects. Vehicles may be difficult to detect because of their location or missing and distorted structural elements, such as license plates and fenders; the appearance of unexpected objects, such as a crowd of pedestrians exiting a venue and crossing a street, animals darting across a roadway, or the chaos that ensues from an incident; inclement weather; multiple vehicles changing lanes; inoperable roadway sensors; high clutter from shrubs, trees, or infrastructure elements such as bridge structural components; and rocks, ice, or snow. Data fusion processes and algorithms that combine data from multiple sensors and other resources offer the potential to enhance traveler safety and mobility, provide higher-confidence vehicle detection and tracking, extend vehicle tracking spatially and temporally, and afford insight into what created the situation causing the observed events and its impact on the occurrence of future events.

The need for real-time and accurate data for implementation of ITS traffic management strategies, including automatic incident detection, active transportation and demand management, route guidance, and work zone, slippery road, and other safety warnings found in connected vehicle applications, underscores the importance of complementary sources of data, i.e., sensors other than those installed in or above the roadway, for traffic flow parameter estimation. The wide spectrum of heterogeneous and independent data sources that can potentially be utilized in a given traffic management situation enable many applications to benefit from sensor and data fusion procedures.[15] Accordingly, data fusion can be applied to produce an improved model or estimate of system parameters or events where the desired model is the state vector describing the traffic phenomena of interest. These estimates may include current or future vehicular speeds, mean speeds, travel times, vehicle classifications, red-light–running assessment, road surface state (i.e., dry, wet, or snow or ice covered), and other information of significance to travelers and traffic management personnel.

2.3 Data Sources for Traffic Management Applications

Innovations in road telematics offer improved methods of traffic data collection. Examples include novel sensor and communications technologies, such as enhanced roadside-mounted sensors that provide multiple data parameters, improved spatial resolution, multilane coverage, or incorporating more than one detection technology; dedicated short-range communications (DSRC) associated with V2V, V2I, and V2P connected and cooperative vehicles;[16,17] in-car sensors that monitor the status of vehicle systems (e.g., braking severity, hazard-warning light activation, velocity, acceleration, steering-wheel position, traction loss, lane departure warning, and windscreen wiper activation); floating car data that provide a traffic management center emissions information in addition to basic traffic flow parameters (volume, occupancy, and speed) linked to the vehicle's location via GPS signals; crowdsourcing data from traveler-initiated inputs (e.g., cellular telephone communications concerning incidents or inclement weather via voice or apps); personal device monitoring that measures the travel behavior of pedestrians, bicyclists, and public transit users; and commercial sources such as SpeedInfo, INRIX, HERE, TomTom International B.V., and Waze.

Collection and analysis of big data from connected vehicles are in their initial stages. Braking severity and hazard warning activation data offer opportunities to identify areas of congestion and incidents on roadways. Vehicle volume and speed data can assist in optimizing delivery van scheduling. Traction loss and windshield wiper activation data can support the issuance of inclement weather and slippery road condition warnings to drivers.

Improvement and expansion of wireless networks and the ubiquitous nature of cellular phone utilization enable location services to exploit cell-phone–equipped drivers as traffic probes.[18,19] Automatic vehicle identification (AVI) systems and technologies—including roadside devices that access toll-tag and trucking industry transponders that transmit clearance credentials to inspection stations, ALPRs, and Bluetooth®-device MAC address readers—support vehicle tracking for roadway design and planning, locating incidents and areas of congestion, and vehicle re-identification for route travel time, speed measurement, and determination of OD pairs. Other types of sensors (not specifically discussed in this book) provide road weather data and weigh-in-motion (WIM) data utilized for road conditioning such as snow and ice removal, fog and slippery road warnings, air quality monitoring, adherence to vehicle weight restrictions, and planning for future road improvements.

Other traditional sources of data continue to play a role in traffic and transportation operations. These include data and information gathered from commercial vehicle radio transmissions, highway service patrol transmissions, emergency call boxes, maps and roadway configuration drawings, toll road and per-lane volume data, planned work zone and sports event schedules, transit system operations, time-of-day and seasonal traffic pattern predictions, and information from other databases.

Basic traffic flow data needed by traffic operations personnel were historically obtained from sensors embedded in the pavement (also referred to as intrusive sensors because they infringe on the roadway pavement) or mounted on roadside poles and signs or overhead structures (also known as nonintrusive sensors). Moreover, sensors may be categorized as active or passive. Active sensors both transmit and receive energy. The received energy is a portion of the transmitted energy that is scattered back into the aperture of the sensor by vehicles, the road, and other objects of interest, and from clutter such as vegetation and roadside structures. Passive sensors only receive energy; i.e., they do not transmit energy of their own. The received energy is a combination of energy emitted and energy reflected into the sensor's aperture by motorized vehicles, bicycles, pedestrians, other objects of interest, the road surface, and road divider structures, and from the surroundings, e.g., trees, buildings, billboard signs, bridges, and tunnel surfaces.

The predominant sensor for obtaining traffic flow data is the inductive loop detector (ILD), which measures temporal traffic flow characteristics at a given location. Other point sensor technologies such as acoustic, ultrasonic, passive infrared, magnetometer, and Doppler microwave can be used to gather roadway network data. While these devices provide point data, they fail in measuring the spatial behavior of traffic flow.[20,21] In addition, their installation and maintenance costs may become comparable to those of more sophisticated sensors when large-area coverage of a roadway network or a multilane intersection is required. Roadside sensors with improved spatial coverage have been developed and deployed to better monitor traffic behavior and support ITS. Among these are visible and infrared spectrum video detection systems and surveillance cameras, and multilane, presence-detecting radar sensors.[22] Table 2.1 summarizes the advantages and limitations of the single and multilane sensors commonly installed on roadways.

Every data source has its unique accuracy, detection area and resolution, and data refresh rate. No one data source may be capable of providing a complete kinematic description of all of the vehicles and pedestrians at every location. Therefore, it is often necessary to gather needed data and information from a variety of devices, which requires that the data be temporally and spatially correlated. To this end, it is valuable to time stamp data since data sources may convey information at different times and intervals. Spatial correlation is needed since different data source modalities may report from different locations or provide information from differently sized detection areas. This may require the transformation of sensor data from each sensor's local spatial coordinate system into a common coordinate system before fusion occurs. Such spatial transformations are often part of sensor registration or alignment procedures, a topic discussed further in Section 2.8.

2.4 Sensor and Data Fusion Architectures

2.4.1 Architecture selection

Architecture selection is dependent on the particular goals and objectives of the sensor and data fusion scenario and application, requirements to incorporate existing agency-owned sensors and other equipment, aesthetic considerations, sensor resolution and output data types, data processing and communication resources (e.g., latency, data rate, and bandwidth requirements), available funding, and the time constraints imposed on development of the system. Goals might include improving spatial and temporal coverage, data accuracy, and operational robustness, i.e., the ability to function under

Table 2.1 Roadway sensor options for traffic management applications.

Sensor	Advantages	Limitations
Inductive loop detector (Active)	Electronics units (detectors) in controller cabinet send an AC current through the wire loop embedded in the roadway. When a metal vehicle passes over the loop, the unit detects the change in inductance of the loop caused by the metal vehicle. Mature, well-understood technology. Excellent counting accuracy when properly installed and maintained. Provides count, presence, and occupancy data. Some models provide classification data.	Not suitable for bridges, or overpasses, viaducts, and poor roadbeds. Reliability and useful life are dependent on installation and maintenance procedures. Installation and maintenance require lane closure. Decreases pavement life. Susceptible to damage by heavy vehicles, road repair, and utilities. Multiple sensors usually required at a site.
Magnetometer (Passive)	Detects perturbation in the Earth's magnetic field created by a metal vehicle. Less susceptible than loops to stresses of traffic. Detects stopped and moving vehicles. Most models transmit data over a wireless RF link. Provides count, presence, and occupancy data.	Small detection zone. Installation requires intrusion into pavement. Installation and maintenance require lane closure. Decrease in pavement life. Multiple sensors usually required at a site.
Magnetic detector (Passive)	Detects same phenomena as above. Can be used where loops are not feasible (e.g., bridge decks). Some models installed under roadway without need for pavement cuts. Less susceptible than loops to stresses of traffic. Provides count and passage data.	Small detection zone. Installation requires pavement cut or tunneling under roadway. Cannot detect stopped vehicles (exception for one model that uses multiple sensors and application-specific software from vendor).
Visible-spectrum video detection system (Passive)	Detects reflected light. Provides best-resolution images. Offers multilane data collection. Provides speed, count, presence, occupancy, limited classification by vehicle length, and other data.	Night operation may require street lights. Affected by clouds, heavy rain and snow, fog, haze, dust, smoke, sun glint, and glare. May be affected by shadows (false or missed calls), or reflections from wet pavement (false calls). Vehicle-to-road contrast, day/night transitions, camera vibration, and debris on camera lens may affect performance. No range data.
Presence-detecting radar (Active)	Detects portion of transmitted energy backscattered into antenna of radar sensor. All weather, day/night operation. Lower frequencies penetrate foliage. Offers multilane data collection.	Vehicle occlusion in distant lanes when side mounted. Offset mounting distance must be accommodated. May require multiple detection zones per lane or other measures to ensure 100% vehicle detection at a stop line.

(continued)

Table 2.1 *(Continued)*

Sensor	Advantages	Limitations
	Provides range, speed, count, presence, occupancy, tracking data, and limited classification by vehicle length.	
Doppler microwave sensor (Active)	Detects portion of transmitted energy backscattered into antenna of sensor. All weather, day/night operation. Provides count, occupancy, and speed of moving vehicles.	Most only detect vehicles traveling greater than some minimum speed; does not detect stopped vehicles or provide presence indication.
Infrared video detection system (Passive)	Detects heat emissions. Fine spatial and spectral resolution imagery. Does not require external illumination. Potential for collecting similar data as a visible-spectrum device. Less affected by rain, fog, haze, dust, smoke, snow, sun glint, glare, and reflections compared to visible-spectrum devices. Day/night operation.	Poor foliage and cloud penetration. Maximization of signal-to-noise ratio (SNR) requires photon detectors that utilize a cooled focal plane (may not be a concern in vehicle or pedestrian detection applications as uncooled pyroelectric detectors are used). Cost may be a concern.
Passive infrared (Passive)	Detects heat emissions. Provides count, presence, occupancy, vehicle class by length, vehicle speed with multiple detection zone sensors. Day/night operation.	Traffic interrupted with overhead installation and repair. Performance possibly degraded by heavy rain, fog, overcast skies, or snow. One sensor per lane required.
Lidar (laser radar) (Active)	Detects portion of transmitted energy backscattered onto a photon detector in the sensor. Fine spatial and spectral resolution imagery. Provides range and reflectance data. Provides velocity, track, count, presence, occupancy, classification data. Day/night operation.	Performance affected when distance from sensor to monitored area is obscured by rain, fog, haze, dust, smoke, or heavy snow. Poor foliage penetration. Most effective when used for monitoring a relatively small area.
Acoustic (Passive)	Detects engine and road noise. Side-mounted, multilane data collection. Provides count, presence, occupancy, and speed. Day/night operation.	May undercount in congested flow. Performance may degrade in rain.
Ultrasonic (Active)	Detects portion of transmitted energy backscattered into sensor from vehicle or road surface. Ease of pole-mounted installation. Provides count, presence, and occupancy. Day/night operation.	Traffic interrupted with overhead installation and repair. Performance affected by variations in temperature and extreme air turbulence. One sensor per lane required. A low pulse repetition frequency device may degrade occupancy measurement on freeways with moderate-to-high-speed vehicles.

changing conditions such as inclement weather, traffic flow variations, and the occurrence of unplanned events. A typical objective may be to process sensor data from diverse and independent sources to provide facility managers and other stakeholders with a complete and coherent understanding of the traffic patterns and events that arise. Other concerns include selecting the location of the data fusion process (for a traffic management application, data fusion may occur at transportation or traffic management centers, freight management offices, first-responder dispatch centers, and in the case of specialized sensors, in the sensor itself); specifying system requirements, including (1) the numbers and types of sensors and other diverse sources of data, (2) communication media required to transmit data to the fusion location or to other sensor subsystems, (3) data storage methods and facilities, (4) additional processing needed to obtain the desired information from the sensor data, and (5) test procedures; and a host of human–computer interaction issues that can be treated as part of Level 5 fusion (see Section 2.12).

The selection of a data fusion architecture also requires an overall system perspective that simultaneously considers the viewpoints of four major participants:[2,23]

1. *System stakeholders*, whose concerns involve system requirements, user desires, operations, and maintenance;
2. *Numerical or statistical specialists*, whose knowledge includes numerical techniques, statistical methods, and algorithm design;
3. *Operations analysts* concerned with the man–machine interface (MMI), transaction analysis, and operational concepts; and
4. *Systems engineers* having expertise in the systems-engineering process, designing and building systems that deliver the required performance, interoperability with other systems, and system integrity. Traffic management personnel at a traffic management center or consultants working for them frequently assume this role in traffic management applications.

2.4.2 Architecture classification

There are several ways to classify sensor and data fusion architectures.[2,3] One taxonomy organizes architectures by where the majority of data processing occurs, e.g., in the individual sensors, distributed at locations where traffic signal controllers are located, or in a central fusion processor at a traffic management center. In another approach, architectures are categorized by the class of data that are combined, e.g., raw data, feature-based data, or decision-type data. Decision-level fusion is optimal for detecting and classifying objects only if the sensors involved rely on independent signature generation phenomena to develop information about the identity of objects in the field of regard. When this is not the case, minimally processed data (such as Level 0 data described in Section 2.6) are sent to a central processor where they are combined using a fusion algorithm capable of detecting, classifying, and tracking objects. Hybrid architectures vary in their makeup and may use combinations of raw, feature, and decision data along with several types of classification algorithms to identify objects. Many hybrid approaches are application specific, such as the multiple-classifier system described in Section 3.9. Disadvantages of hybrid fusion include increased processing complexity and possibly increased data transmission rates.

Hybrid fusion can manifest in the form of hierarchical and distributed architectures. Neyman–Pearson and Bayesian formulations of the distributed sensor detection problem for parallel, serial, and tree data fusion topologies can be found in Viswanathan and Varshney.[24] A hierarchical architecture contains fusion nodes arranged such that the lowest-level nodes process sensor data and send the results to higher-level nodes to be combined. Figure 2.2 shows one example of a distributed fusion architecture where the fusion nodes process data from multiple sensors and send their results to other fusion nodes for further processing with additional sensor data or to the consumers of the fusion information.[25] Fixed superior–subordinate relationships do not exist in a fully distributed architecture. Each node can communicate with other nodes subject to connectivity constraints. The communication can be adaptive and dependent on the information content and requirements of the individual nodes. Significant savings in communication resources are achieved when the higher-level nodes collect processing results periodically. The advantages of a distributed fusion architecture and the issues raised through its use are summarized in Table 2.2.

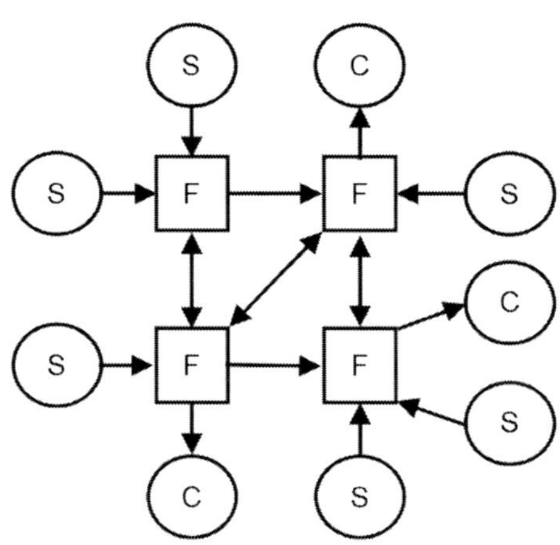

S = sensor or information source
C = information consumer
F = fusion node

Figure 2.2 Distributed fusion architecture.

Table 2.2 Advantages and issues associated with distributed fusion architecture.

Advantages	Issues
Lighter processing load at each fusion node because processing is distributed over multiple nodes. No requirement to maintain a large centralized database since each node can be linked to its own database. Reduced communication load because data are not sent to and from a central processing site. Faster access to fusion results due to reduced communication delay. Increased survivability due to elimination of single-point failure mode (a flaw in a centralized fusion architecture).	Architecture: sharing of fusion responsibility among nodes, e.g., identification of sensors or sources reporting to each node and objects of interest for which each node is responsible. Communications: connectivity and bandwidth of the nodal communication network, identification of information sources and sinks, and establishing need for raw data or processing results for each node. Algorithms: methods used by nodes to efficiently and effectively fuse data and provide situation awareness (i.e., who, when, what, and how).

Dasarathy[26] discusses other architecture frameworks, such as those based on the category of the entities at both the input and output of the fusion system (e.g., data, features, or decisions), and the notion of the fusion architecture being constructed to provide feedback to the individual sensors.

Additional fusion frameworks, including those developed for robotics, may be applicable to the automated driving scenarios being pursued by the U.S. in its Connected Vehicle Program and in other countries where vehicle-to-everything (V2X) communications will become more prevalent in the decades to come. Durrant-Whyte and Henderson[27] classify robotics architectures according to selections made from among four independent design dimensions: (1) centralized–decentralized, (2) local–global interaction of components, (3) modular–monolithic, and (4) heterarchical–hierarchical. The most prevalent combinations are: centralized, local interaction, and hierarchical; decentralized, global interaction, and heterarchical; decentralized, local interaction, and hierarchical; and decentralized, local interaction, and heterarchical. Each of the following architectures may contain meta, algorithmic, conceptual, logical, and execution architectures within them:

- The *meta architecture* is a set of high-level considerations that strongly characterize the system structure. For example, a decentralized, local-interaction, heterarchical architecture has a meta architecture characterized as follows. It is decentralized, meaning that no component is central to the operation of the system and that the communication is peer to peer. Also, it does not contain central facilities or services (e.g., for communication or name and service lookup or timing). These features create scalable, fault-tolerant, reconfigurable systems. Local interactions imply that the number of communication links does not change with the network size. Moreover, the number of messages also remains constant, making the system scalable and reconfigurable as well. Modularity leads to interoperability as derived from interface protocols, reconfigurability, and fault tolerance where failure may be confined to individual modules.

- The *algorithmic architecture* is a specific set of information fusion and decision-making methods. An example of an algorithmic architecture is one employing three fusion concepts: belief fusion, utility fusion, and policy selection. Belief fusion is achieved by communicating all beliefs (defined as a probability distribution of the state space that applies to the scenario of interest) to neighboring platforms. Utility fusion is implemented "by separating the individual platform's partial utility into the team utility of belief quality and local utilities of action and communication. The downside is that the potential coupling between individual actions and messages is ignored because the utilities of action and communication remain local."[27] The communication and action policies are chosen by maximizing expected values.
- The *conceptual architecture* contains the granularity and functional roles of components such as mappings from algorithmic elements to functional structures. The data types for the algorithmic architecture described above include beliefs related to the objects or events in the scenario of interest, future plans as related to the beliefs of interest, and future actions related to the beliefs of interest.
- The *logical architecture* is designed from the conceptual architecture, and consists of detailed canonical component types (i.e., object-oriented specifications) and interfaces to formalize intercomponent services.
- The *execution architecture* defines the mapping of logical components to execution (runtime) elements such as processes and shared libraries. The execution architecture contains a deployment view that shows the mapping of physical components onto the nodes of the physical system and a source code view that explains how the software implementing the system is organized.

2.5 Detection, Classification, and Identification of a Vehicle

An observer's ability to interpret and extract information from an image is a function of the contrast and brightness of the object (in traffic management applications, the object is a vehicle or conveyance of some type, a pedestrian, or a fixed object such as a lamp or signal support structure, sign bridge, building, or tree), properties of the device that enhance the image or the algorithm that processes the data, and the physiological response characteristic of the human eye–brain interface when a human interprets the imagery. Contrast is determined by the image signal-to-noise ratio (SNR), while brightness is controlled by the luminance of the object of interest. A review of the literature concerning human interpretation of images may be found in Ref. 28.

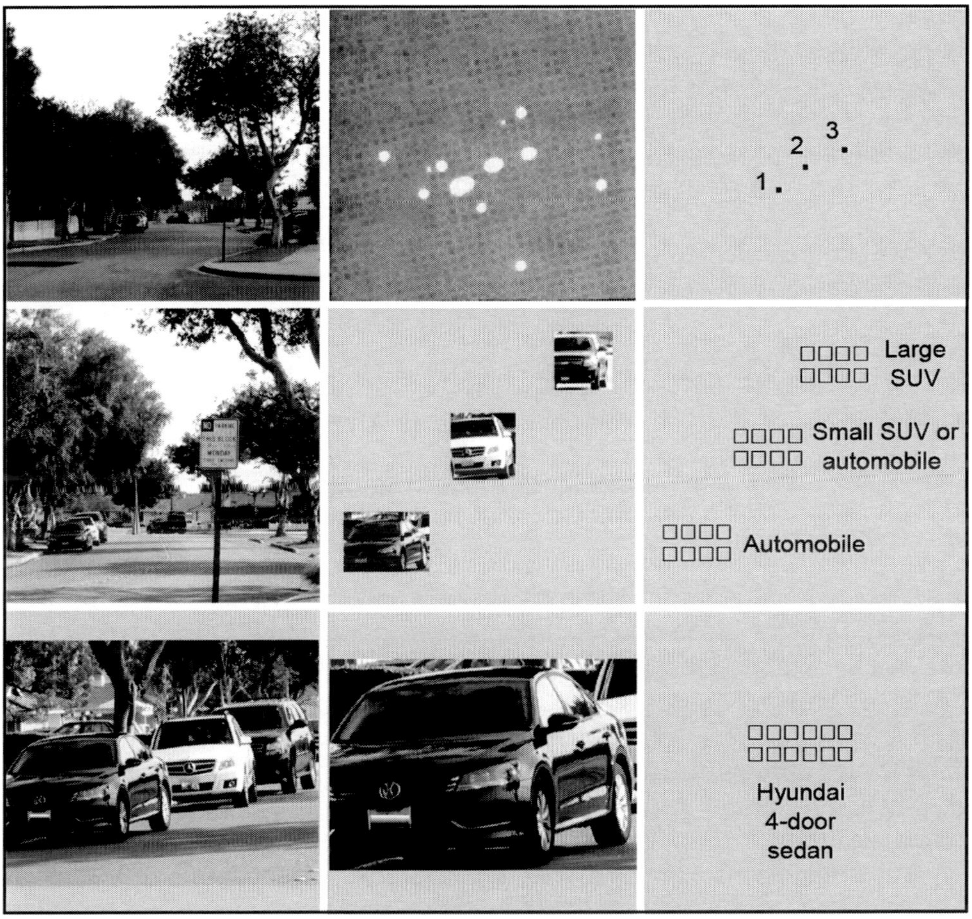

Figure 2.3 Detection, classification, and identification of a vehicle. The squares in the last column of rows two and three indicate the approximate number of pixels needed across the minimum dimension of an object to achieve a 50% probability of classifying or identifying the object, respectively, assuming an adequate SNR (between 2.5 and 3 approximately), a one-dimensional (1D) detector array, and a scene-scanning mechanism. Classification and identification probabilities for 2D arrays are based on the number of pixels resolved within the boundaries of the object rather than across its minimum dimension. The 2D arrays require only approximately 75% of the pixels needed by the 1D arrays to achieve the desired probability.[29]

Detection, classification, and identification are terms often associated with locating objects and with data fusion. The first row in Fig. 2.3 depicts the detection process, whose purpose is to determine if an object of interest is present. This process involves differentiating between images or data that belong to relevant objects (e.g., the parked cars shown in the first column and indicated by the numbers 1, 2, and 3 in the last column of the row) and those that belong to clutter or objects not of interest, such as trees, rocks, fences and other barriers, lamp poles, or perhaps puddles of water. The second row shows the classification process, namely, it discerns the class to which an object belongs. For traffic management, classes of interest include sedans, buses and other transit vehicles, emergency vehicles, sport utility vehicles, pickup trucks, 18-wheelers and other truck configurations, pedestrians, and bicycles, for example. The third row shows the identification process, which describes the objects to the limit of the observer's or data-processing algorithm's knowledge, e.g., identifying the manufacturer and model of a vehicle.

2.6 The JDL and DFIG Data Fusion Models

The motivation for creating the original JDL data fusion model[6] is summarized by Llinas et al.[7] in the following statement: "Much of its value derives from the fact that identified fusion functions have been recognizable to human beings as a 'model' of functions they were performing in their own minds when organizing and fusing data and information. It is important to keep this 'human centric' sense of fusion functionality, since it allows the model to bridge between the operational fusion community, the theoreticians, and the system developers." The data processing or function levels in the model are based on a partitioning scheme that accounts for the combined and interdependent effects of varying levels of abstraction and problem-space complexity. Subsequent revision to the model produced the six-level version shown in Fig. 2.4.[8,9] Although the user refinement level was not officially recognized until 2013 in Ref. 30, it is indicated here to reflect its current status.

The augmented DFIG model seen in Fig. 2.5 supports exploitation functions and information management for situation awareness, massive data analytics for contextual awareness, and domain-specific needs for mission

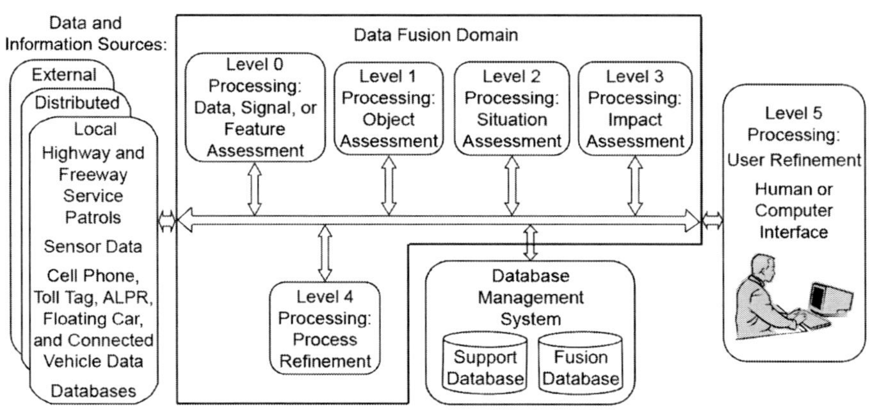

Figure 2.4 Six-level JDL data-fusion–processing model. Data fusion levels are not a prescription for designing systems. Instead, they are meant to be a convenient way to categorize the functions that occur in a comprehensive data fusion process.

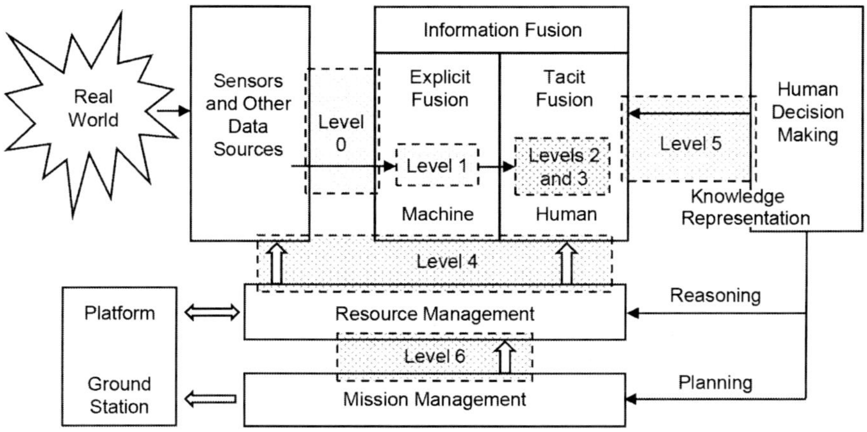

Figure 2.5 DFIG information fusion model.

awareness. The resulting information fusion model separates the data fusion and management functions (namely, resource management and mission management), broadens the roles of information fusion to include Level 0 and a new Level 6 (mission management), aligns similar functions (namely, Levels 2 and 3 concerning abstract information such as threat, intent, and goals that can be treated as part of the larger situation awareness function), subsumes process refinement (Level 4 processing) into resource management, and acknowledges the physical, cognitive, and social analyses a human performs as an element of information fusion (user refinement or Level 5 processing).[30,31] Situation awareness and its decomposition into situation assessment and impact and threat assessment are expanded on in Section 2.10.

Resource management functions are partitioned into sensor control, platform placement, and user selection to meet mission objectives. Sensor control functions are based on spatial, temporal, and spectral preferences. Temporal requirements are centered on the user's need for timely information to initiate actions, while spatial needs are established from the mission goals. Level 6 or mission management is a sub-element of the larger systems management function. It is concerned with adaptive spatiotemporal control of assets (e.g., in the traffic management setting, these may include roadside sensor and moveable sign placement, floating car and roving freeway service patrol deployment, and traffic adaptive signal control implementation where beneficial), route planning (e.g., as it affects deployment of first responders or service to new and growing communities), and goal setting to support team decision making and actions that recognize social, economic, political, and organizational constraints. The DFIG model acknowledges that situation assessment includes tacit functions that are inferred from Level 1 explicit representations of object assessment. Since the unobserved aspects of the situation assessment problem cannot be processed by a computer, user knowledge and reasoning are required to manage these.[30]

The terms and examples below, which describe several of the processing levels in the JDL fusion model, reflect the connection of this book to ITS and, therefore, may differ from the usual parlance found in the military literature:

- Level 0 processing (referred to as preprocessing, data assessment, or signal or feature assessment) transforms the data generated by the contributing source into a format that is more effectively processed by the user. Typical operations at this level are spatiotemporal data and image registration, noise removal, normalizing, formatting, ordering, batching, and compressing input data.[5–7,32–36] Level 0 processing may even identify sub-objects or features in the data that are used later in Level 1 processing.
- Level 1 processing (object assessment or entity assessment) exploits identity and state estimation data fusion algorithms to identify and track objects (e.g., vehicles moving and stopped, vehicle platoons, pedestrians, bicyclists, and animals) and events (e.g., incidents and crowds of pedestrians at a crosswalk or jay walking). For traffic management, the data of interest arise from sources such as those listed in Section 2.3.
- Level 2 processing (situation assessment) combines the results of Level 1 processing with information from pertinent databases to develop a description of current relationships or patterns among objects and events as framed by the environment of interest and to indicate the potential onset of an event, response plans, and inferences about the distribution of the objects identified in Level 1 processing.
- Level 3 processing (impact assessment) evaluates the situation (event) and traffic flow patterns and other data to indicate the likely effect of a traffic event on traffic flow, such as timing of the occurrence of the event (e.g., traffic congestion, incident, fire, or police action) and duration of the event. A significant amount of information from external databases is usually needed to support the Level 2 and Level 3 fusion processes.
- Level 4 processing (process refinement), an element of resource management, seeks to improve the entire data fusion process by continuously refining predictions and assessments, and evaluating the need for additional sources of data. Level 4 processing is alternatively viewed as an element of resource management as discussed in Section 2.11.
- Level 5 processing (user refinement) is concerned with enabling a human to access, interpret, and apply the results of the fusion process.[37–40] The DFIG model approaches user refinement as part of the broader knowledge management task.[31]
- Level 6 processing (mission management), an element of systems management, involves adaptive spatiotemporal control of assets, route

planning, and goal determination to support team decision making and actions that recognize social, economic, political, and organizational constraints.

Figure 2.6 illustrates the functions of the Level 1 through Level 3 fusion processes. In some applications, object detection, classification, and state estimation occur simultaneously rather than in separate paths as shown in the figure.

Low-level processes (Level 1 as shown here, although Level 0 processing is also included in the low-level category) support object classification, identification, and tracking using data that describe object locations, kinematics, and attribute types. Level 1 fusion analyzes and combines the data or imagery from each sensor or source concerning objects in the scene of interest to enhance and update detection, classification, and tracking of the objects of interest. The algorithms and operations employed at this level include physical models, syntactic analysis, parametric and nonparametric methods, cognitive-based approaches, and correlation and association of data and tracks.

The Level 2 and Level 3 high-level functional processes support situation and impact assessment usually with abstract symbolic information concerning threats, intents, and goals.[41,42] These processes interpret the environment, recognize behavior patterns, associate entities and events, categorize the situations causing the observed events, and predict future behavior.

An important caveat in the design of data fusion systems is that the data fusion levels just described are intended only as a convenient categorization of data fusion functions. Data fusion levels were never intended to be, nor should they be taken as, a prescription for designing systems: do Level 0 fusion first,

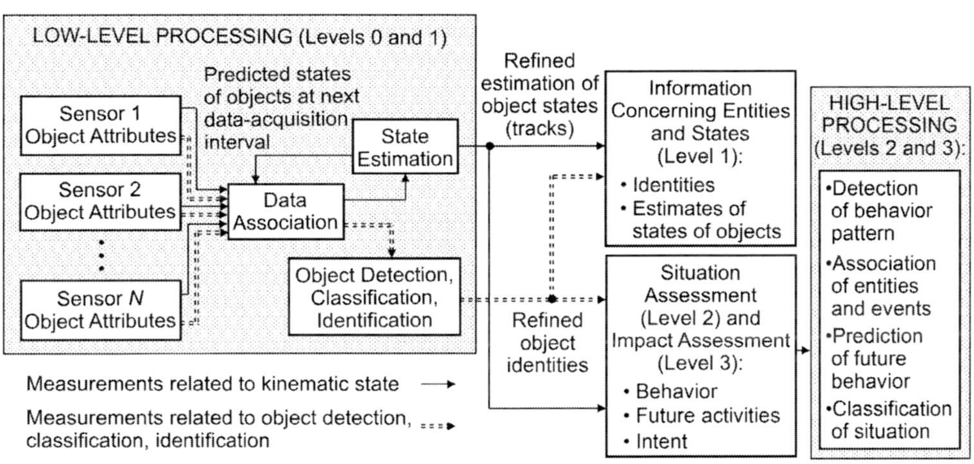

Figure 2.6 Data fusion processing Levels 0, 1, 2, and 3 of the JDL model (adapted from Ref. 5).

then Level 1, then Level 2, etc. According to Hall and Steinberg, "the model was intended for communications among data fusion researchers and implementation engineers, rather than a prescription for implementing a fusion system or an exhaustive enumeration of fusion functions and techniques."[43] Processing should be partitioned in terms of the individual system requirements and to provide outputs that satisfy the needs of the decision maker.[44]

In the traffic management community, data collection, situation and impact assessment, and performance refinement most likely are performed by the organization responsible for managing traffic on a particular facility, although functions such as data collection may be assigned to commercial sources. For example, if roadside sensor, MAC address, and cell phone incident-location data are collected, traffic management agency personnel most likely have a lead role in these data fusion processes. The data may be subject to preprocessing or may pass directly to one of the other fusion levels. In the future, as connected and self-driving vehicles become commonplace on roadways, companies providing access to rideshare assistance and Mobility as a Service (MaaS) may also become involved in sensor and data fusion operations.

Level 1 fusion is expanded on in Sections 2.7 and 2.8, while Levels 2–5 are discussed further in Sections 2.10–2.12.

2.7 Level 1 Fusion: Detection, Classification, and Identification Algorithms

Figure 2.7 contains a taxonomy of Level 1 detection, classification, and identification algorithms.[5,32,34,45,46] The major algorithm categories are

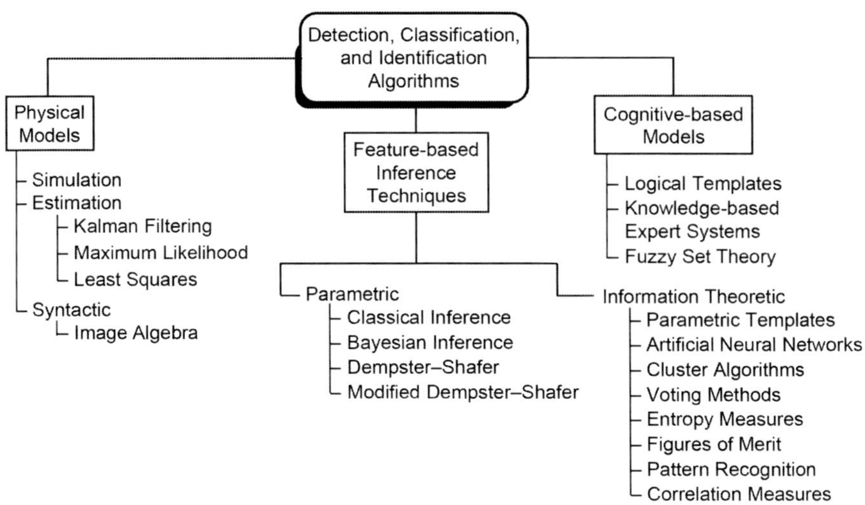

Figure 2.7 Detection, classification, and identification algorithm taxonomy for Level 1 processing.

physical models, feature-based inference techniques, and cognitive-based models. Other mathematical concepts (not shown in the figure) are also utilized for Level 1 data fusion. These include random set theory, conditional algebra, and relational event algebra. Random set theory deals with random variables that are sets rather than points. Goodman, Mahler, and Nguyen use random set theory to reformulate multisensor, multi-object estimation problems into single-sensor, single-object problems.[47] These authors also apply the theory to incorporate ambiguous evidence (e.g., natural language reports and rules) and various expert system methods (e.g., fuzzy logic and rule-based inference) into multisensor, multi-object estimation. Conditional-event algebra is a type of probabilistic calculus suited for contingency problems such as knowledge-based rules and contingent decision making. Relational-event algebra is a generalization of conditional-event algebra that provides a systematic basis for solving problems involving pooling of evidence. Still other data fusion approaches combine several of the illustrated methods, such as combinations of Bayesian logic with fuzzy logic, Dempster–Shafer theory with fuzzy logic, artificial neural networks with fuzzy logic, and artificial neural networks with Dempster–Shafer theory.

The applications that appear with the algorithm descriptions in this section are examples that may not involve data fusion but do demonstrate their pertinence to the world of ITS and traffic management when applied to process data from a single source. Applications that appear later in Section 2.13 are those that do involve multiple sensor and data fusion examples from ITS.

2.7.1 Physical models

Physical models replicate object discriminators that are easily and accurately observable or calculable. Examples of discriminators include geometrical features such as edges, lines, line widths, line relationships (e.g., parallel, perpendicular), arcs, circles, conic shapes, and size of enclosed area; structural features represented by surface area, relative orientation, orientation in vertical and horizontal ground plane, and juxtaposition of planes, cylinders, and cones; radar cross-section as a function of aspect angle; infrared emissions as a function of vehicle type, engine temperature, or surface characteristics such as roughness, emissivity, and temperature; multispectral signatures; and height profile images.[34]

This approach classifies and identifies objects by matching the prestored model feature data to observed data as illustrated in Fig. 2.8. Signatures or imagery of objects gathered by a sensor are analyzed for pre-identified physical characteristics or attributes that are input into an identity declaration process. Here, the characteristics identified by the analysis of the observed data are compared with stored physical models containing features characteristic of the objects of interest and other items that may be present,

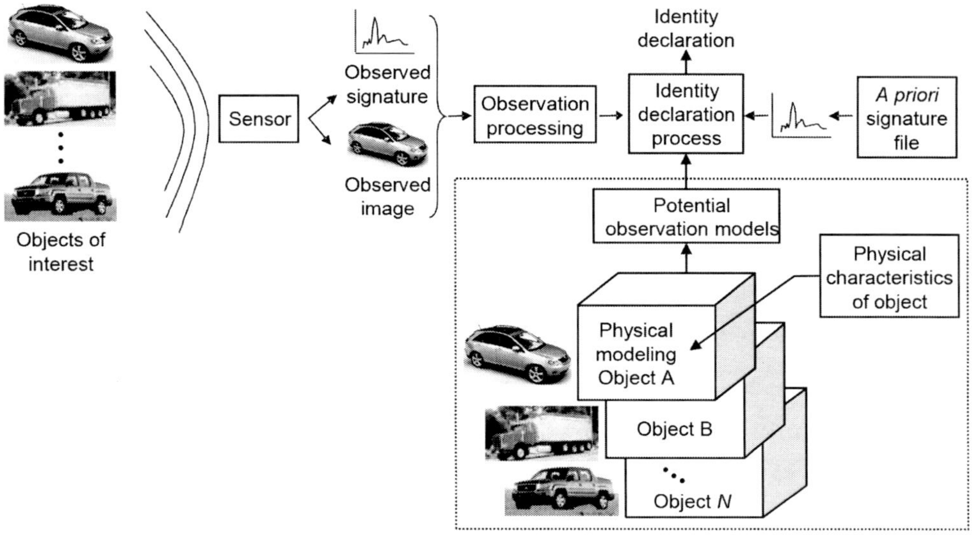

Figure 2.8 Physical modeling concept.

such as trees, sign posts, or buildings. The stored model having the closest match to the real-time sensor data is declared to be the correct identity of the object in the sensor's field of view.

Physical modeling techniques include simulation, estimation, and syntactic methods. Simulation is used when the physical characteristics to be observed can be accurately and predictably modeled. Estimation processes include Kalman filtering, maximum likelihood, and least squares approximation. The Kalman filter provides a general solution to the recursive, minimum mean-square state estimation problem within the class of *linear* estimators. It minimizes the mean-squared error as long as the tracked object's dynamics and measurement noise are accurately modeled. The Kalman filter and its nonlinear-motion counterparts are examples of physical models since the kinematics of the tracked objects are modeled.

The *extended Kalman filter* (EKF) is applied when nonlinearities are present in the observation matrix or covariance matrices of the process and measurement noise sources. The EKF linearizes about the current mean and covariance of the state using first-order Taylor approximations to the time-varying transition and observation matrices. There are alternatives to the EKF when the system of interest is highly nonlinear.[3,10,11] The first of these employs an unscented Kalman filter (UKF), which operates on the premise that it is easier to approximate a Gaussian distribution than it is to approximate an arbitrary nonlinear function. Instead of linearizing using Jacobian matrices as in the EKF, the UKF uses a deterministic sampling approach to capture estimates of the mean and covariance with a minimal set of carefully chosen sample points. Application of the UKF allows the posterior mean and covariance to be accurately represented to the third order (Taylor series expansion) for any nonlinearity. The EKF, in contrast, only achieves first-order accuracy. The computational complexity of the UKF is on the order of that of the EKF. An application of the UKF to speed estimation is found in Ye, Zhang, and Middleton.[48]

Monte Carlo techniques are appropriate for problems where state transition models and measurement models are highly nonlinear. These techniques are flexible as they do not make any assumptions regarding the probability densities to be approximated. They describe probability distributions as a set of weighted samples of an underlying state space. The samples simulate probabilistic inference usually by applying Bayes' theorem. Many simulations are performed and, by analyzing the statistics of the simulations, a probabilistic picture of the process can be discerned. Sequential Monte Carlo filtering is a simulation of the recursive Bayes' update equations using sample support values and weights to describe the underlying probability distributions.

Particle filters extend the sequential Monte Carlo algorithm by utilizing a weighted ensemble of randomly drawn samples, called particles, as an approximation of the probability density of interest. When applied within a Bayesian framework, particle filters approximate the posterior probability of the system state as a weighted sum of random samples. The random samples are usually drawn (predicted) from the prior density (transition model) with their weights updated according to the likelihood of the given measurement (measurement model). An example of the application of particle filtering to the estimation of traffic flow speeds and travel times is given by Cheng, Qiu, and Ran.[49]

Syntactic methods, although listed under physical models, appear again as part of pattern recognition, a subset of information theoretic techniques. Syntactic pattern recognition is applied when the significant information in a pattern is not merely the presence or absence of numerical values, but rather the interconnections of features that yield its structure. Pattern similarity is assessed by quantifying and extracting structural information, utilizing, for example, the syntax of a formally defined language. Typically, syntactic approaches formulate hierarchical descriptions of complex patterns from simpler subpatterns or primitives.

2.7.2 Feature-based inference techniques

Feature-based inference techniques perform classification or identification by mapping data, such as statistical knowledge about an object or recognition of object features, into a declaration of identity. Feature-based algorithms may be further divided into parametric and information theoretic techniques (i.e., algorithms that have some commonality with information theory).

2.7.2.1 Parametric techniques

Parametric classification directly maps parametric data (e.g., features) into a declaration of identity. Stochastic properties of features may be modeled, although physical models are not used. Parametric techniques include classical inference, Bayesian inference, Dempster–Shafer evidential reasoning, and modified Dempster–Shafer methods.

While not truly a data fusion method, classical inference is mentioned briefly because Bayesian inference and Dempster–Shafer evidential theory can be thought of as generalizations of this technique. *Classical inference* may be utilized to find the probability that an observation can be attributed to the presence of an object or event, given an assumed hypothesis. Its major disadvantages are: (1) difficulty in obtaining the density function that describes the observable used to classify the object, (2) complexities that arise when multivariate data are encountered, (3) the capability to assess only two hypotheses at a time, and (4) the inability to take direct advantage of *a priori* and likelihood probabilities.

Bayesian inference resolves some of the difficulties with classical inference. It updates the *a priori* probability of a hypothesis given a previous likelihood estimate and additional observations, and is applicable when more than two hypotheses are to be assessed.[34,50] The limitations of Bayesian inference include: (1) difficulties in defining the prior probabilities and likelihood functions, (2) complexities that arise when multiple potential hypotheses and multiple conditionally dependent events are evaluated, (3) the mutual exclusivity required of competing hypotheses, (4) the need for conditionally independent data sources, and (5) the inability to account for general uncertainty. When multiple sensors collect data and apply Bayesian inference to determine the identity of an object, each sensor provides an identity declaration D or hypothesis about the object's identity based on the observations and a sensor-specific algorithm. The previously established performance characteristics of each sensor's classification algorithm (developed either theoretically or experimentally) provide estimates of the likelihood function, i.e., the probability $P(D|O_i)$, that the sensor will declare the object to be of a certain type, given that the object is, in fact, type i. The sensor declarations are then combined using a generalization of Bayes' rule to produce an updated, joint probability for each entity O_i.

Dempster–Shafer evidential reasoning generalizes Bayesian inference to allow for uncertainty by distributing support for a proposition (e.g., that an object is of a particular type) not only to the proposition itself, but also to the union of propositions (disjunctions) that include it, and to the negation of a proposition. Any support that cannot be *directly* assigned by an information source to the propositions or their negations is allocated to the set of all propositions that form the hypothesis space, called the frame of discernment. This corresponds to assigning probability mass to the uncertainty class.

Support provided by multiple sensors for a proposition is combined using Dempster's rule. The sensors must act as independent sources when offering support for a proposition. Bayesian and Dempster–Shafer techniques produce identical results when all propositions are singleton (i.e., consist of only one object) and are mutually exclusive, and there is no support assigned to uncertainty. Limitations of the method include the inability to make direct use of prior probabilities when they are known, and the counterintuitive output sometimes produced when support for conflicting propositions is large. Another issue concerns the selection of a method to calculate the probability mass that defines the sensor's support for a proposition. This topic is discussed further in Section 4.7. Several modifications to Dempster's rule have been proposed that use probability transformations to better accommodate conflicting beliefs[51] and, in some cases, prior knowledge and spatial information.[52–58]

2.7.2.2 Information theoretic techniques

Information theoretic techniques transform or map parametric data into an identity declaration. They relate a similarity in identity to a similarity in the observable parameters. No attempt is made to directly model the stochastic aspects of the observables. The techniques in this category are parametric templates, artificial neural networks, cluster algorithms, voting methods, entropy-measuring procedures, figures of merit, pattern recognition, and correlation measures.

In *parametric templating*, single and multisensor or multispectral data acquired over time are matched with preselected conditions to determine if the observations contain evidence to identify an entity. Templating can be applied to object identification, event detection, and situation assessment.[5,35] Figure 2.9 shows an application of parametric templating to the identification of a vehicle's class using time-domain signatures obtained from high-sampling-rate ILDs.[59] The normalized signature magnitudes and difference in the magnitudes of consecutive points are overlaid on a template such as the one depicted in the lower right portion of the figure. Identification is made when the parameters lie in a region that corresponds to the signature of a known vehicle class. If the parameters fall into a region that has not been previously mapped, then the vehicle class is declared to be unknown. In contrast with the truck signatures depicted in the figure, passenger cars typically produce a unimodal, symmetrical waveform.

Artificial neural networks (ANNs) are hardware or software systems that are trained to map input data into selected output categories. The transformation of the input data into output classifications is performed by artificial neurons that attempt to emulate the complex, nonlinear, and massively parallel computing processes that occur in biological nervous systems.[60]

The particular ANN in Fig. 2.10 is an example of a fully connected, three-layer feedforward network. In a fully connected network, each processing

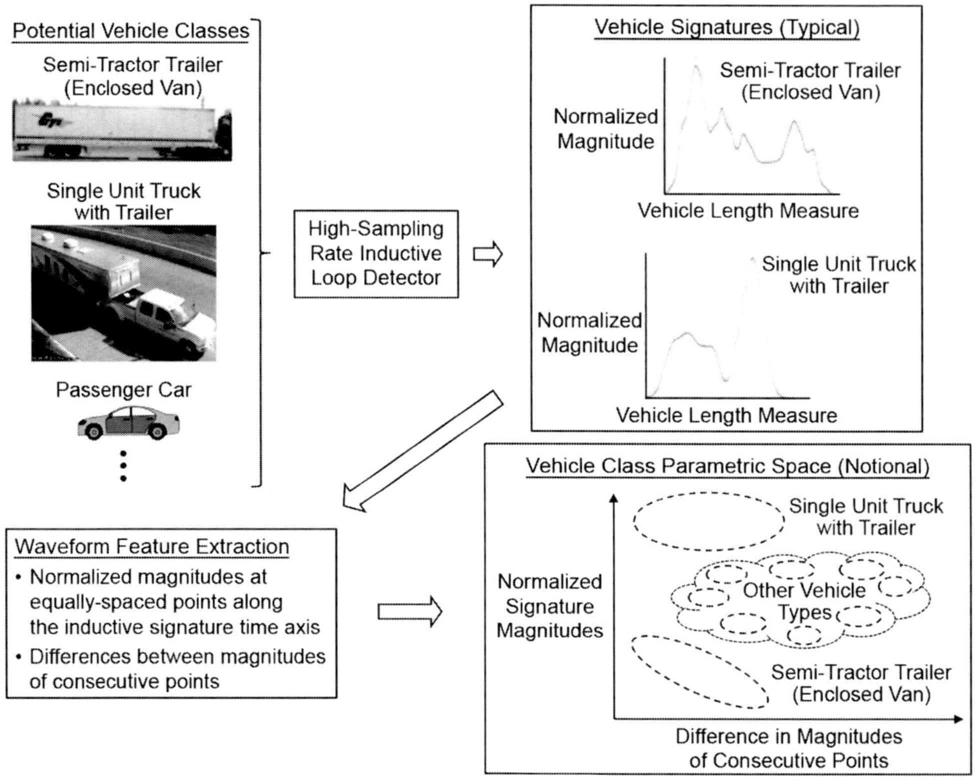

Figure 2.9 Parametric templating concept (the truck signatures in top right of the figure are from Ref. 59).

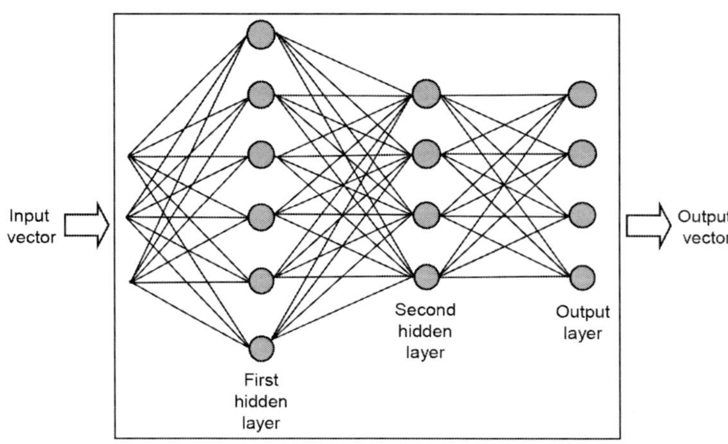

Figure 2.10 Fully connected, feedforward artificial neural network.

element receives inputs from every output in the preceding layer. In general, a feedforward network contains a hierarchy of processing elements that are organized as a series of two or more mutually exclusive layers. The input elements are a holding place for the values applied to the network. These elements do not implement a separate mapping or conversion of input data, and their weights are insignificant. The last, or output, layer permits the final state of the network to be read. Between these two extremes are zero or more layers of hidden elements. The hidden layers remap or scale the inputs and results of previous layers' analyses using weights and, thereby, produce a more separable or more easily classifiable representation of the data. Other forms of feedforward networks exist, such as one where the processing elements receive signals directly from each input component and from the output of each preceding processing element.

Traffic management applications of ANNs include a study by Wang and Kim[61] that applies a fully connected, feedforward ANN to the prediction of crashes. Their model classifies predictions into one of two binary outcomes: incident or no incident. The model uses a set of link traffic speeds across the network to estimate the probability of a particular outcome within a certain time window in the future at a given location. This large-scale ANN consists of one input layer, five hidden layers, and one output layer. The input layer contains $744 \times N$ input variables representing speed observations from the entire study network of 744 detector links over N previous time slices. The output layer contains a single output variable representing the probability of crash occurrence in the given target cell at a predefined future time interval.

The probabilistic neural network (PNN) illustrated in Fig. 2.11 is another form of artificial neural network utilized in ITS applications that differentiates

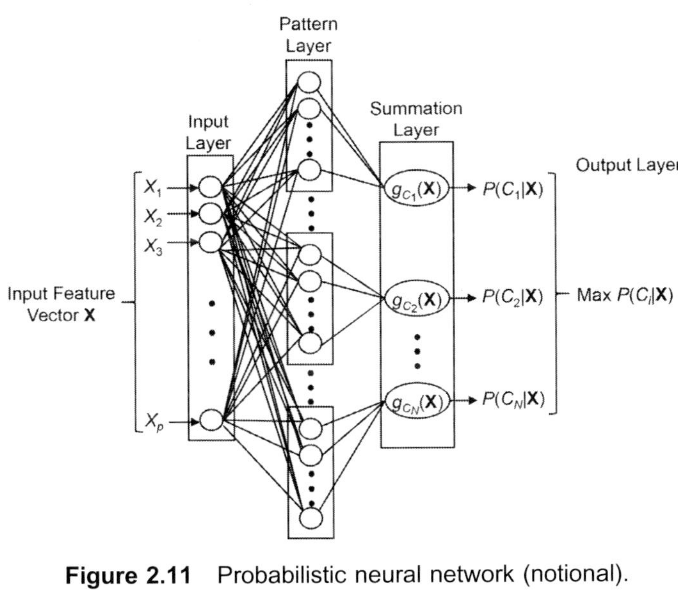

Figure 2.11 Probabilistic neural network (notional).

among classes of objects or events. For example, Oh, Oh, and Ritchie[62] used a PNN for hazardous traffic condition warning, as did Oh and Ritchie[63] for vehicle classification, Hernandez, Tok, and Ritchie[64] for inclusion in a multiple-classifier system for commercial vehicles, and Ahmed, Banik, Hasnat, et al.[65] for transit service quality ranking.

The major advantages of a PNN are that training requires only a single pass, and decision surfaces are guaranteed to approach the Bayes optimal decision boundaries as the number of training samples grows. Moreover, the shape of the decision surfaces can be made as complex or as simple as desired by choosing the appropriate value for a smoothing parameter that appears in the Parzen estimator of the probability density functions (pdfs) of the different classes to be separated.[66,67] The Parzen estimators or kernels enable the PNN to instantiate a multivariate Bayesian classifier. Kernel functions $k(u)$ are smooth and continuous and have the property that

$$\int_{\infty}^{\infty} k(u)du = 1, \tag{2.1}$$

allowing them to be used as a pdf when the kernel is non-negative, i.e., $k(u) \geq 0$ for all u. Commonly used Parzen kernel distributions include the uniform, biweight, triangular, Epanechnikov (parabolic), and Gaussian.[68,69]

The four layers in a PNN are the input layer, pattern layer, summation layer, and output layer. The number of nodes in the input layer equals the number of input features to the neural network. The input layer distributes the normalized input values to all of the neurons in the pattern layer. The pattern units first form a dot product of the input vector $\mathbf{X}$ with a weight vector $\mathbf{W}_i$ such that $\mathbf{Z}_i = \mathbf{X}\mathbf{W}_i$. Next, the pattern units execute a nonlinear operation on $\mathbf{Z}_i$ before outputting their activation levels to the summation layer. Instead of the sigmoid activation function commonly associated with multilayer feedforward neural networks and the backpropagation algorithm, the nonlinear function used in a PNN is $\exp[(\mathbf{Z}_i - 1)/\sigma^2]$. When $\mathbf{X}$ and $\mathbf{W}_i$ are normalized to a unit length, the nonlinear function becomes $\exp\left[-\frac{(\mathbf{W}_i-\mathbf{X})^{\mathrm{T}}(\mathbf{W}_i-\mathbf{X})}{2\sigma^2}\right]$. The network is trained by setting the $\mathbf{W}_i$ weight vector in one of the pattern units equal to each $\mathbf{X}$ pattern in the training set, and then connecting the output of the pattern unit to the appropriate summation unit. A separate neuron (pattern unit) is required for every training pattern.

When a global, single smoothing parameter and a Gaussian pdf are employed, the Parzen kernel estimator becomes

$$g(\mathbf{X}) = \frac{1}{(2\pi)^{p/2} n\sigma^p} \sum_{i=1}^{n} \exp - \frac{\|\mathbf{X} - \mathbf{X}_i\|^2}{2\sigma^2}, \tag{2.2}$$

where $\mathbf{X}$ and $\mathbf{X}_i$ are the vectors of input variables and the ith training vector, respectively; σ is the smoothing parameter equal to the standard deviation of the input variables; and p is the dimension of the input feature space.

The summation layer contains several nodes that represent the pattern classes C_i of interest. Its output is the posterior probability for each class. The single-node output layer reveals the pattern class for a given input feature vector by implementing the Bayesian decision strategy of choosing the class that gives the largest value of the posterior probability. Training of the PNN is relatively fast as each input is shown to the network only once. Users can add new training data without retraining the network. Unlike traditional neural networks, no learning rule is required to train a PNN, and no predefined convergence criteria are needed.

Table 2.3 lists the operating principles and applications of additional types of ANNs.[2] Included are the multilayer perceptron, the radial basis function network, Kohonen self-organizing network, Grossberg adaptive-resonance network, counterpropagation network, and Hopfield network. The Kohonen, Grossberg, and counterpropagation networks use unsupervised learning. Supervised learning utilizes pattern class information introduced to the network through training samples; unsupervised learning does not.

Cluster algorithms group data into natural sets or clusters that are interpreted by an analyst to see if they represent a meaningful object category. All cluster algorithms require a similarity metric or association measure that describes the closeness between any two feature vectors, e.g., one that represents the input data and one that represents a potential class to which the data belong.

These algorithms operate with five basic steps: (1) selection of sample data, (2) definition of the set of variables or features that characterize the entities in the sample, (3) computation of the similarities among the data, (4) use of a cluster analysis method to create groups of similar entities based on data similarities (see Waltz and Llinas[5] Chapter 7 for a summary of clustering methods and the algorithms they employ), and (5) validation of the resulting cluster solution. The application of cluster algorithms may lead to biased results because of the heuristic nature of these algorithms. In general, data scaling, choice of similarity metric and algorithm, and sometimes even the order of the input data may substantially affect the resulting clusters. Hence, application of cluster methods must be judged on their effectiveness and ability to form consistent and meaningful identity clusters.[5,34]

Figure 2.12 depicts one representation of cluster analysis. Observations or data acquisition from known objects are gathered during a training cycle, followed by identification and extraction of features that assist in uniquely classifying the objects of interest. A feature-based classifier operates on the feature vector $\mathbf{Y}$ and allocates specific regions in the feature space to those objects. When training is complete, unknown objects are observed and the

Table 2.3 Descriptions of artificial neural networks.[2]

Type	Key Operating Principles	Applications
• Multilayer perceptron	• A multilayer feedforward network. • Uses signum (hard limit) or sigmoid (soft limit) threshold nonlinearities. • Trained with supervised learning. • Errors are minimized using backpropagation algorithm to update weights applied to the input data by the hidden and output network layers. • No more than three layers are required because a three-layer perceptron can generate arbitrary complex decision regions. • Number of weights equals the number of hidden-layer neurons.	• Accommodates complex decision regions in the feature space. • Target classification. • Speech synthesis. • Nonlinear regression.
• Radial basis function network	• Uses a multilayer feedforward network to provide regularization, i.e., a stabilized solution using a non-negative function to embed prior information (e.g., training examples that provide smoothness constraints on the input–output mapping), which converts an ill-posed problem into a well-posed problem. • Minimizes a cost function that is proportional to the difference between the desired and actual network responses. • Approximate solutions for the cost function utilize a number of basis functions that is less than the number of input data points to reduce computational complexity. • One form of radial basis function network gives its response as a linear superposition of the products of weights and multivariate Gaussian basis functions with centers located at the input data points, and widths equal to the standard deviation of the data. • Other basis functions typically used are piecewise-linear approximation, cubic approximation, the multiquadratic function, and the inverse-multiquadratic function. • Trained with supervised learning.	• Target classification. • Image processing. • Speech recognition. • Time series analysis. • Adaptive equalization to reduce effects of imperfections in communications channels. • Radar point source location. • Medical diagnosis.
• Kohonen self-organizing network	• Feedforward network that uses an unsupervised learning paradigm (processes unlabeled data, i.e., data where the desired classification is unknown). • Utilizes a mathematical transformation to convert input data vectors into output graphs, maps, or clustering diagrams.	• Speech recognition.

(*continued*)

Table 2.3 (*Continued*)

Type	Key Operating Principles	Applications
• Kohonen self-organizing network (continued)	• Individual neural-network clusters self-organize to reflect input pattern similarity. • Output nodes are extensively interconnected with many local connections. • Overall structure of the network can be regarded as an array of matched filters that competitively adjust input element weights based on current weights and goodness of match of the output to the training set input. • Trained with winner-take-all algorithm. The winning node is rewarded with a weight adjustment, while the weights of the other nodes are unaffected. Winning node is the one whose output cluster most closely matches the input. • Network can also be trained with multiple-winner unsupervised learning where the K neurons best matching the input vector are allowed to have their weights adjusted. The outputs of the winning neurons can be adjusted to sum to unity.	
• Grossberg adaptive-resonance network	• Unsupervised learning paradigm that employs feedforward and feedback computations. • Teaches itself new categories and continues storing information without rejecting pieces of information that are temporarily useless, as they may be needed later. Pattern or feature information is stored in clusters. • Uses two layers—an input layer and an output layer. The output layer itself has two sublayers: a comparison layer for short-term memory and a recognition layer for long-term memory.	• Pattern recognition. • Target classification.
• Counter-propagation network	• Consists of two layers that map input data vectors into bipolar binary responses (–1, +1). Allows propagation from the input to a classified output, as well as propagation in the reverse direction. • First layer is a Kohonen layer trained in unsupervised winner-take-all mode. The first layer organizes data, allowing faster training to perform associative mapping. The outputs of the Kohonen layer neurons are binary unipolar values 0 and 1. Input vectors belonging to the same cluster activate the same neuron in the Kohonen layer. • Second layer is a Grossberg layer that orders the mapping of the input vectors into the bipolar binary outputs. The result is a network that behaves as a lookup memory table.	• Target classification. • Pattern mapping and association. • Data compression.

(*continued*)

Table 2.3 *(Continued)*

Type	Key Operating Principles	Applications
• Hopfield network (a type of associative memory network)	• Associative memory networks belong to a class of neural networks that learn according to a specific recording algorithm. They usually require *a priori* information, and their connectivity (weight) matrices are frequently formed in advance. Trained with supervised learning. • Writing into memory produces changes in neural interconnections. Transformation of the input signals by the network allows information to be stored in memory for later output. • No usable addressing scheme exists in associative memory since all memory information is spatially distributed and superimposed throughout the network. • All neurons are connected to each other. • Network convergence is relatively insensitive to the fraction of elements (15 to 100%) updated at each step. • Each node receives inputs that are processed through a hard limiter. The outputs of the nodes (±1) are multiplied by the weight assigned to the nodes connected by the weight. • The minimum number of nodes is seven times the number of memories stored.	• Problems with binary inputs. • Data association. • Optimization problems. • Optical character recognition.

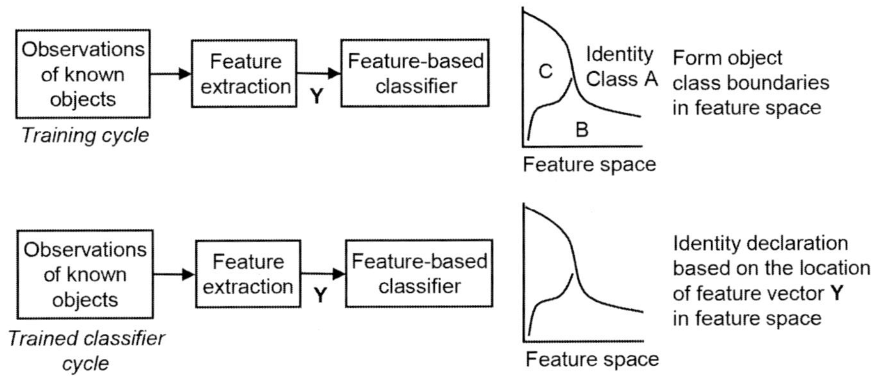

Figure 2.12 Cluster analysis framework (adapted from Ref. 34).

same features are extracted from their signatures. The feature-based classifier then identifies the region in the feature space that best corresponds to the feature vector obtained from the unknown object.

Traffic management applications of cluster algorithms include research by Kim and Jang,[70] who explored spatiotemporal profiles of crash events, i.e., a scatter plot of crashes versus postmile and time of day, to acquire features symptomatic of crashes along a section of urban freeway. The features were developed from plots of the cumulative number of crashes versus postmile number. This procedure requires a large number of samples to acquire a statistically meaningful clustering pattern. Plots indicative of crashes indicate an increased crash rate along the highway, i.e., an increase in the slope of the plot, followed by a decrease. However, if the data collection interval is too long, it obscures the spatial cluster's time-varying patterns. A 1-hr interval was found to give satisfactory results. Crash rates intensified during the rush hour peak and, as traffic congestion became aggravated, tended to increase more drastically at specific sections of the roadway such as those in close proximity to ramp junctions.

Allahviranloo, Regue, and Recker[71] predicted activity or travel patterns (i.e., type, duration, scheduling, and location) of individuals with differing socio-demographic characteristics by clustering their activity patterns using affinity propagation and *k*-means clustering algorithms. Distances between pairs of activity patterns representing 10-min time intervals were computed using the weighted sum of sequential alignment and agenda dissimilarity distance measures. Clustering was implemented using affinity propagation to generate initial centroids for the *k*-means clustering algorithm that followed.

Rinaldi and Tampère[72] applied clustering for anticipatory network traffic control. They recognized that disturbances to the network's conditions from a control action, such as a change in the green times at a traffic signal intersection, a variation in a ramp-metering rate, or a change in variable speed limits, cause the users to react to the new cost conditions according to user equilibrium principles that seek to minimize the total time spent in the network. The reaction is twofold: (1) in terms of propagation, a change in control might influence a bottleneck, e.g., by causing a reduction of flow downstream of the bottleneck itself and creating a congestion wave that spills backwards upstream; and (2) in terms of route choice behavior when a sufficiently large change in cost over a given route might trigger a diversion to a new route with a lower cost in terms of reduced travel time. This dynamic clustering algorithm relies on a heuristic that tracks fluctuations in the second-order derivative of the link speed vector. The rationale for this choice is that variations in either route choice behavior or congestion wave propagation will be reflected in link state changes. If the traffic regime transitions from free flow into congestion, or *vice versa*, a new clustering of the flow-control Jacobian matrix representing the network may be needed.

Ohn Bar[73] studied vehicle detection centered on object subcategories formed from clusters of visual and fine-grained 3D geometrical features. The clustering algorithms produced a predetermined number of clusters. Visual clustering relied on shape features deduced from pixels in LUV color space, gradient magnitudes, and gradient orientations to capture vehicle shapes. Geometrical features were obtained from 3D orientation, position, occlusion level and type, and other features suggestive of the vehicle's shape. Typical features included a histogram of oriented gradients (HOG), aspect ratio, configuration and occlusion type (not occluded, partially occluded, heavily occluded, or unknown occlusion type), occlusion level (percentage of the vehicle outside the camera field of view), 3D orientation, and geometrical and visual features. Each feature sample was assigned a color according to its orientation and occlusion level. Four clustering strategies were investigated. Strategy 1 employed a uniform binning of orientation and occlusion levels. This supervised approach encodes the prior information based on orientation and occlusion level as these account for the majority of appearance variations. Strategy 2 formed clusters over 3D geometry features directly using *k*-means or spectral unsupervised clustering. This strategy encompasses all of the variables that influence appearance variations (such as occlusion statistics, level, and type) and appeared to perform well. Strategy 3 used 3D geometry features to initialize a visual subcategorization routine, such as a latent support vector machine or other framework. Fusion of the geometric and visual features is significant because a partition in geometrical space may not be correlated with a partition in visual space. This scheme produced minor improvements to the final detection results in the experiments. Strategy 4 clustered only visual features without using geometrical data. Strategies 1 and 2 performed best in this study.

Zhao, Xu, Wu, and Liu[74] investigated a combination of clustering as a source of features and a backpropagation feedforward ANN as a classifier to identify pedestrians and vehicles from lidar data. Clustering was performed with a revised density-based spatial clustering of applications with noise

(DBSCAN) algorithm. Three features were extracted from each cluster to serve as inputs to the ANN classifier. The features were total number of points (vehicle clusters have more points than pedestrian clusters located at the same distance from the lidar sensor because vehicles are larger in size), 2D distance of the mean center of each cluster to the lidar sensor, and the direction of the cluster point distribution (the distribution of pedestrian cluster points is generally along the vertical direction, while the distribution of vehicle cluster points is along the horizontal direction). The accuracy of the clustering and classification processes was 93.6%. Three effects were observed to cause failure of the process, namely, receiving a small number of lidar signals when pedestrians and vehicles were far from the sensor; blocking of part of a vehicle or pedestrian by other pedestrians, vehicles, or facilities; and having several pedestrians located close to each other but far from the sensor.

Voting methods combine detection and classification data from multiple sensors by treating each sensor's declaration as a vote in which majority, plurality, or decision-tree rules are applied, often with the aid of Boolean algebra. Additional discrimination can be introduced via weighting of the sensor's declaration if the sensor-system architect harbors doubts concerning a sensor's reliability in specific environments or the independence of its output from that of other sensors. The voting technique is most effective when the sensors act independently of each other in providing outputs that signify the presence of an object or event of interest. The sensor outputs also contain information that expresses the sensor's confidence that its declaration is correct and information concerning the object's location. Detection modes consisting of combinations of declarations from two and three sensors have been proposed to provide robust performance in clutter, inclement weather, and environments where a particular sensor's signature may be suppressed, e.g., by countermeasures in a military application. The detection modes utilize sensor confidence levels appropriate to the number of sensors in the detection mode. For example, the confidence levels for sensors in a two-sensor detection mode are typically greater than those in a three-sensor mode. A Boolean-algebra–based voting algorithm can be derived to give closed-form expressions for the multiple-sensor system's estimation of the object's detection probability and false-alarm probability.[2]

An example of a three-sensor system having four detection modes ($A_1B_1C_1$, A_2C_2, B_2C_2, and A_3B_3) is shown in Fig. 2.13. The three sensors are denoted by A, B, and C, and sensor confidence levels are denoted by subscripts 1, 2, and 3, where 1 represents the lowest confidence level of a sensor and 3 the highest. Only two- and three-sensor detection modes are considered in this example. The AND gates combine the sensor confidence level outputs corresponding to each detection mode of the system. The OR gate confirms an object or event detection by accepting the decision of the activated detection mode.

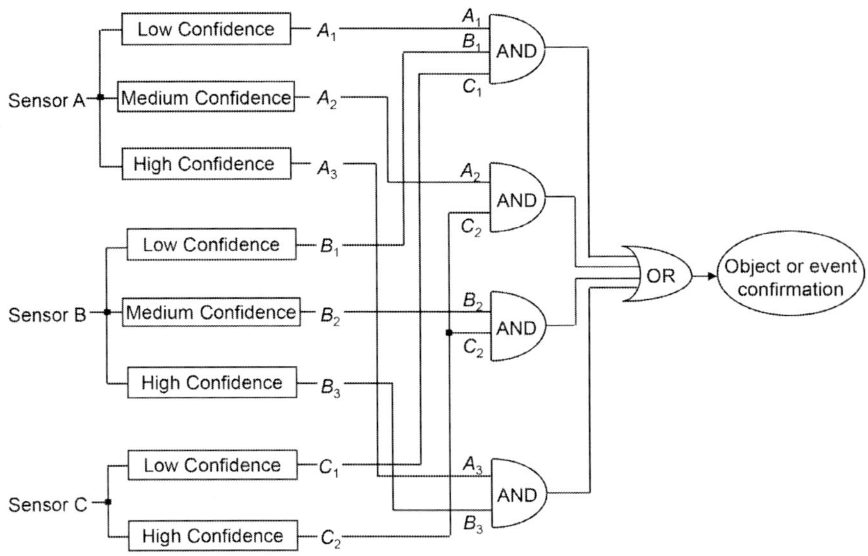

Figure 2.13 Voting fusion with a three-sensor system having multiple-sensor detection modes. Confidence levels are represented by the outputs of sensors A, B, and C through the subscripts 1, 2, and 3 that indicate low, medium, and high confidence level outputs, respectively.

Pattern recognition concerns the description or classification of data. The three major approaches to pattern recognition are statistical (or decision theoretic), syntactic (or structural), and ANNs. In statistical pattern recognition, a set of characteristic measurements or features are extracted from the input data and used to assign the feature vector to one of c classes. Assuming that features are generated by a state of nature, the underlying statistical model represents a state of nature, a set of probabilities, or probability density functions that correspond to a particular class.[75] Syntactic pattern recognition is applied when the significant information in a pattern is not merely the presence or absence of numerical values, but rather the interconnections of features that yield structural information. The structural similarity of patterns is assessed by quantifying and extracting structural information using, e.g., the syntax of a formally defined language. Typically, syntactic approaches formulate hierarchical descriptions of complex patterns from simpler subpatterns or primitives. Neural computing was described earlier in this section. Descriptions of the other algorithms in the information theoretic category are found in Klein[2] and Waltz and Llinas.[5]

2.7.3 Cognitive-based models

Cognitive-based models, including logical templates, knowledge-based systems, and fuzzy set theory, attempt to emulate and automate the decision-making processes employed by human analysts.

2.7.3.1 Logical templates

Templating matches predetermined and stored patterns against observed data to infer the identity of an object or to assess a situation. Parametric templates that compare real-time patterns to stored patterns can be combined with logical templates derived, for example, from Boolean relationships.[2,5] Fuzzy logic may also be applied to the pattern-matching technique to account for uncertainty in either the observed data or the logical relationships that define a pattern.

2.7.3.2 Knowledge-based expert systems

Knowledge-based systems incorporate rules and other knowledge from known experts to automate the object identification process. These systems retain the expert knowledge for use at a time when the human inference source is no longer available. Computer-based expert systems frequently consist of four components: (1) a knowledge base that contains facts, algorithms, and a representation of heuristic rules; (2) a global database that contains dynamic input data or imagery; (3) a control structure or inference engine; and (4) a human–machine interface. The inference engine processes the data by searching the knowledge base and applying the facts, algorithms, and rules to the input data. The output of the process is a set of suggested actions that is offered to the end user.[76]

The knowledge-based system in Fig. 2.14 depicts processed sensor waveform data or imagery as the source of the features that identifies the object or situation. Three types of rules are listed to assist in correlating information contained in the real-time feature vector with information in the stored knowledge base. *Syntactic rules* are expressed as IF–THEN statements. The IF or antecedent clause states the conditions that must be present for the

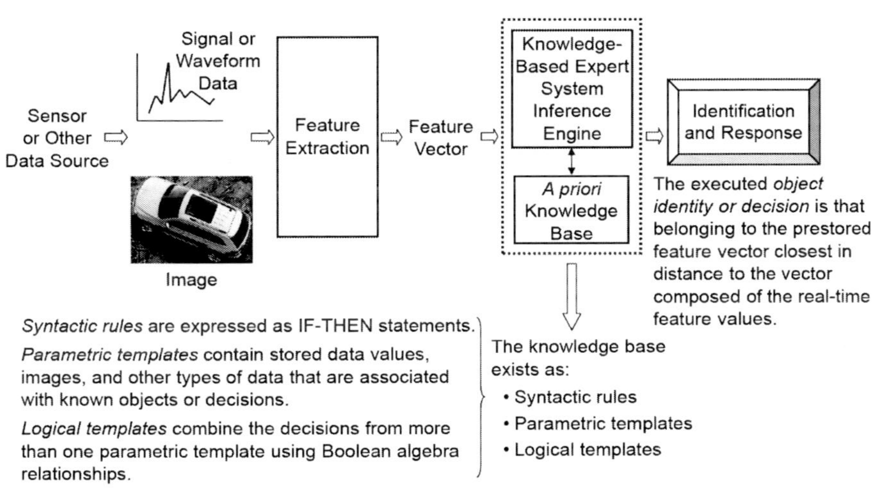

Figure 2.14 Knowledge-based expert system concept.

action specified in the THEN or conditional clause to occur. Expert systems typically rely on binary on–off logic and probability to develop the inferences found in the IF–THEN statements. *Parametric templates* contain stored data values, images, and other types of information that are associated with known objects or decisions. *Logical templates* combine the decisions from more than one parametric template using Boolean algebra relationships. The executed object identity or decision is the one belonging to the prestored feature vector closest in distance to the vector composed of the real-time feature values.

2.7.3.3 Fuzzy set theory

Fuzzy set theory or fuzzy logic opens the world of imprecise knowledge or indistinct boundary definition to mathematical treatment. It facilitates the mapping of system state-variable data into control, classification, or other outputs.[77–79] There are four essential elements to a fuzzy system, namely, fuzzy sets, membership functions, production rules, and a defuzzification mechanism.

Fuzzy sets are the state variables defined in imprecise terms. In Fig. 2.15, the sets represent height categories of people. In conventional set theory, the set boundaries leave no doubt as to the person's height class. However, in fuzzy set theory, the set boundaries are vague; i.e., they overlap into neighboring sets. Therefore, fuzzy set theory allows a person to belong to more than one fuzzy set simultaneously.

Membership functions are the graphical representation of the boundary between fuzzy sets. Membership can range from 0 (definitely not a member) to 1 (definitely a member). The activation value equals the value of the membership function at which it is intersected by the input variable at the time instant being evaluated.

Production rules (also known as fuzzy associative memory), specified in the form of IF–THEN statements, govern the behavior of the system. They are the mechanism that assigns the membership value of an output state variable in a given fuzzy set. The antecedent or condition block of the rule

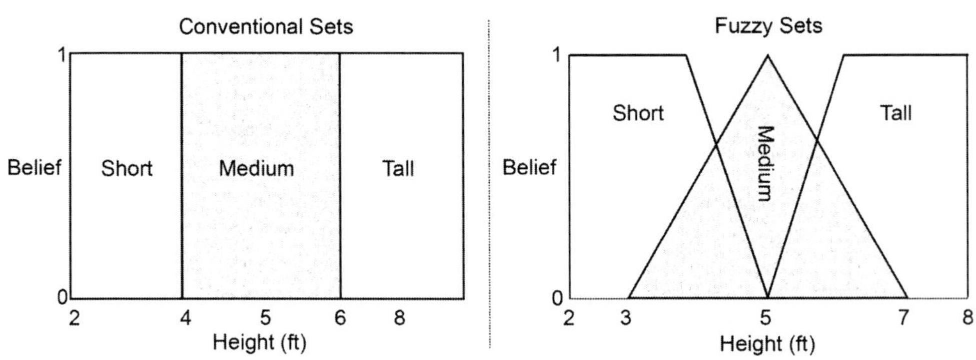

Figure 2.15 Conventional (left) and fuzzy (right) sets representing a person's height.

begins with IF, and the consequent or conclusion block begins with THEN. The value assigned to the consequent block equals the *logical product* of the activation values of the antecedent membership functions. If the antecedent block for a particular rule is a compound statement connected by AND, the logical product is the *minimum* value of the corresponding activation values of the antecedent membership functions. If the antecedent block for a particular rule is a compound statement connected by OR, the logical product is the *maximum* value of the activation values. An expert defines the production rules and fuzzy set boundary shapes that represent the characteristics of each input and output variable. Often the number of production rules is not unique; i.e., more than one collection of production rules may apply to the problem. If this is the case, it is common to select the scheme containing the minimum number of rules.

Defuzzification is the process that converts the result of the application of the production rules into the crisp output value used to control the system. Defuzzification may be implemented in several ways. Most applications execute a center-of-mass (COM) or fuzzy centroid computation, although other methods are available. These include sum of center (SOC), which is less mathematically complex than the COM calculation; height maximum (HM), which offers reduced computational complexity compared to COM and SOC because the areas of the membership functions under the output fuzzy set are not computed; mean of maxima (MOM); first of maxima (FOM); last of maxima (LoM); smallest of maximum (SOM); largest of maxima (LOM); and bisector of area (BOA), which divides the total area into two regions of equal area.

Fuzzy set theory is intuitively appealing in that it permits uncertainties in knowledge or identity boundaries to be applied to such diverse applications as identification of battlefield threats, tracking of objects by means of a fuzzy Kalman filter, and control of industrial and automotive processes.[2] Unlike neural networks that sum throughputs, fuzzy systems sum outputs.

2.8 Level 1 Fusion: State Estimation and Tracking Algorithms

Figure 2.16 displays a taxonomy for state estimation and tracking algorithms used in Level 1 processing.[5,32,34,45] At the top level, the figure shows the options available for (1) conducting data-driven or track-driven searches to update tracks and (2) correlating and associating data and tracks. Correlation and association are further separated into data alignment; data and track association; and position, kinematic, and attribute estimation. The majority of the algorithms are concerned with data and track association techniques.

In data-driven systems, data consisting of combinations of range, azimuth, elevation, and range-rate sensor measurements (assuming a 3D radar as the sensor) initiate a search through the known track file for *tracks*

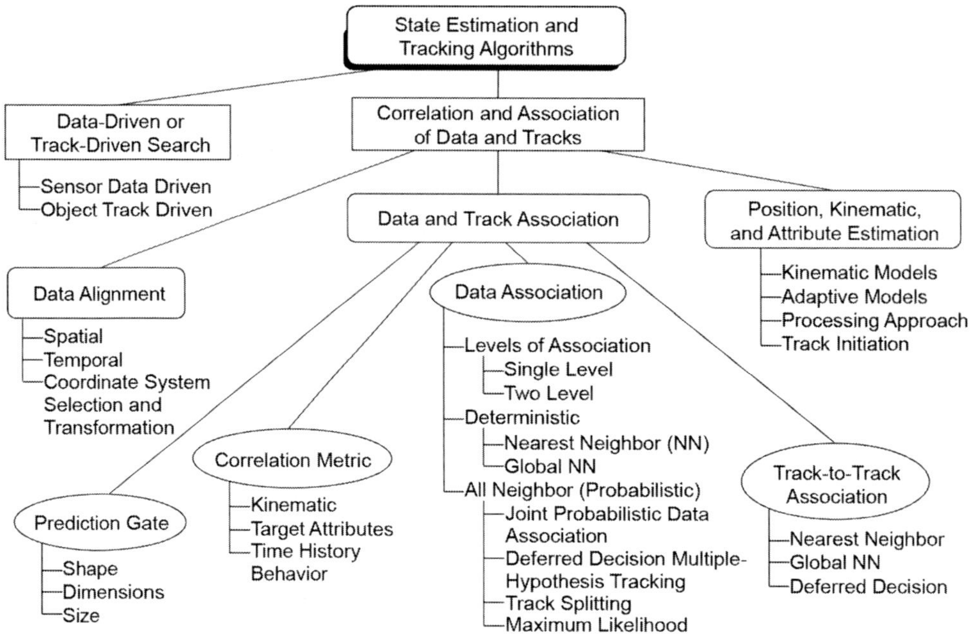

Figure 2.16 State estimation and tracking algorithm taxonomy for Level 1 processing.

that can be associated with the measurements. Track-driven systems use a primary sensor for tracking and use the object track to direct other sensors to acquire data or search databases for *measurements that can be associated with particular tracks.* Tracks are hypothetical constructs in a computer that estimate the position, velocity, and acceleration of the objects of interest given a time-ordered sequence of measurements. Each track ideally represents a unique physical object or entity (not clutter). Other measures of track quality are continuity and accuracy. Continuity denotes continuous motion of an object without jumps or gaps over time. Accuracy, along with stability, is a measure that express the adequacy of the tracks for the intended application.

The proper correlation and association of measurement data and tracks from multisensor inputs ultimately generate optimal central track files. Correlation and association require procedures that assist data alignment, prediction gates, correlation metrics, data and track association methods, and position, kinematic, and attribute estimation. Whenever multiple sensors are detecting multiple objects, there is a need to not only register the data, but also to correlate the measurements with the individual objects of interest and then associate the measurement data and objects to update the state estimate.[34,80] Errors introduced by measurement inaccuracies, coordinate transformations, and unknown object motion are accounted for through the data alignment process. Sensor registration removes sensor position and velocity measurement bias. This process usually begins with the specification of an origin from which the data from different sensors can be referenced and eventually combined. It also attempts to remove or account for calibration bias errors that occur at the individual sensor nodes. Bias errors arise from differences in the sensor range, azimuth, elevation, and time offsets, as well as the coordinate conversion process itself.[81,82]

Data correlation and data association assist in selecting and combining multisource data to update or create object tracks, i.e., trajectories. Correlation determines which sensor measurements to pair with each moving object of interest. Association is the process of updating the object state (i.e., position and velocity) and predicting its future position at the time of the next measurement. Association may also create a new object track if the data do not appear to correlate with any existing object track. The Kalman filter, its variants (extended and unscented Kalman filters), Monte Carlo techniques, and particle filtering are techniques discussed in Section 2.7.1 that are utilized to perform this step. These processes are explained in detail in Bar-Shalom and Fortmann,[80] Bar-Shalom, Li, and Kirubarajan,[83] Li and Jilkov,[84–90] as well as in the references therein.

It is instructive here to further describe the data and track correlation and association processes and to briefly discuss several of the data association methods so that the reader will gain an appreciation of what can often be a complex and perhaps misunderstood process when applied to ITS. Techniques used for position, kinematic, and attribute estimation can be found in Klein[2] and its references.

2.8.1 Prediction gates, correlation metrics, and data association

A *prediction gate* is a mechanism that controls the association of data sets into one of two categories, namely, candidates for track update or initial observations for forming a new tentative track. The gate size is selected to obtain a balance between maximizing detection probability (use of large-sized gates) and minimizing misassociation probability (use of small-sized gates). The size of the gates reflects the combination of calculated or otherwise anticipated errors in predicting the object's position and velocity at the next measurement opportunity, sensor measurement errors, and desired probability of correct association. Figure 2.17 illustrates this concept. Establishing the gate size and predicting an object's location at the next measurement opportunity may be less complicated for the ITS problem than for the aircraft-tracking problem for which it was originally developed since vehicles usually travel on known roadway lanes. Methods are also required to address how data appearing in more than one gate and missing data will be treated.

Correlation metrics quantify the closeness of measurement data to existing tracks. They also appear in track-to-track association to help associate tracks produced by different sensors. In general, metrics are evaluated using the kinematic parameters (e.g., range, range rate, angle, and position) and object attributes (e.g., temperature, size, shape, and edge structure) that are observed

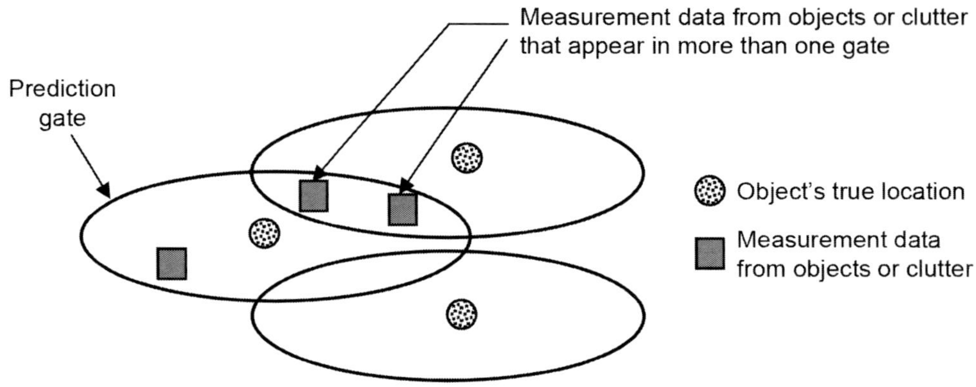

Figure 2.17 Data association as aided by prediction gates.

and measured. The metric can be based on geometric distance (e.g., Euclidean distance) or statistical measures of correlation between observations and predictions (e.g., Mahalanobis distance), heuristic functions such as figures-of-merit that use the kinematic and object attribute information, and measures that quantify the realism of an observation or track based on prior assumptions such as track lengths, object densities, or track behavior.[5,34]

In a multiple-object and multiple-sensor scenario, *data association* refers to the statistical decision process that associates sets of measurement data from overlapping gates, multiple returns in a gate, clutter in a gate, and new objects that appear in a gate on successive scans for the purpose of updating existing tracks or initiating new tracks. Thus, data association partitions the measurements into sets that could have originated from the same objects.[24]

2.8.2 Single- and two-level data and track association

Association techniques that merge data and tracks from several sources into a single track usually employ either single-level tracking systems or two-level tracking systems. In a single-level tracking system, measurement data from several sensors are transmitted to a single processing node. Here the data are correlated and associated to initiate new tracks and update estimates of existing tracks in the central track file.

Two-level tracking systems have four variants: (1) performing track-to-track association at the sensors and at a central node, (2) using sensor data to form tracks locally that are then sent to a central node for track association, (3) transmitting sensor data or tracks to a central node for track association, and (4) executing sensor track association locally and at a central node. The first type of two-level tracking system maintains separate sensor-level and central-level trackers. Each sensor-level tracker initiates and updates tracks based on its own data. The tracks generated by each sensor are sent to a central site where track-to-track association is performed and redundant tracks are eliminated, producing a central track file. The second two-level system forms tracks with local measurement data only. The resulting tracks are reported to a designated track management center for distribution to the users. The third uses either sensor measurement data or sensor tracks to originate and maintain a central track file. Track-to-track association of sensor tracks is initiated at a central node to form a central track file. The fourth two-level system distributes all correlated measurement data to all tracking subsystems for association with new or existing tracks. This approach forms tracks with all available data processed identically at all sensor subsystems, creating a common representation at each site.

2.8.3 Deterministic and probabilistic (all-neighbor) association

In general, there are two distinct approaches to the data association problem, deterministic and probabilistic or all neighbor. The simpler deterministic method includes nearest neighbor and global nearest neighbor data association. It takes the most likely of several possible associations and completely ignores the possibility that this selected association may be inappropriate. The alternatives are probabilistic approaches based on a Bayesian framework (probabilistic data association and joint probabilistic data association), multiple-hypothesis tracking, and maximum likelihood. The probabilistic data association methods are applied in situations where a measurement may fall inside the intersection of two or more validation gates of several different objects of interest (and, therefore, could have originated from any of these objects or from clutter) and when multiple returns from a large object or a closely spaced group of objects occur.[80] Multiple-hypothesis tracking allows the association of data to more than one track until a definitive assignment can be made at a later time.[82] A variation of multiple-hypothesis tracking, called track splitting, associates each measurement in the gate with a track but does not specifically generate new tracks, nor does it compute the probability of correct association. Maximum likelihood selects the most likely single set of measurement data for association with a track.

2.9 Data Fusion Algorithm Selection

Choosing a data fusion algorithm requires the same overall system perspective that was described in Section 2.4 and takes into consideration the concerns and expertise of the system stakeholders, numerical or statistical specialists, operations analysts, and systems engineers.

How does one know which data fusion algorithm or technique to use in a given application? A starting point is to base the choice on the degree to which the technique makes correct inferences and the availability of required computer resources and algorithm input parameters.[2,3] The selection process also seeks to identify algorithms that meet the following goals:

1. Maximum effectiveness: Algorithms are sought that make inferences with maximum specificity in the presence of uncertain or missing data. Required *a priori* data such as likelihood functions and probability masses are often unavailable for a particular scenario and must be estimated within time and budget constraints.
2. Operational constraints: The selection process should consider the constraints and perspectives of both automatic data processing and the analyst's desire for tools and useful products that are executable within the time constraints posed by the application. If the output products are examined by more than one decision maker, then multiple sets of user expectations must be addressed.
3. Resource efficiency: Algorithm operation should minimize the use of computer resources (when they are scarce or in demand by other processes), e.g., CPU time and required input and output devices.
4. Operational flexibility: Evaluation of algorithms should include the potential for different operational needs or system applications, particularly for data-driven algorithms versus alternative logic approaches. The ability to accommodate different sensors or sensor technologies may also be a requirement in some systems.
5. Functional growth: Data flow, interfaces, and algorithms must accommodate increased functionality as the system evolves.

Many of the Level 1 entity assessment data fusion algorithms are mature in the context of mathematical development. They encompass a broad range from numerical techniques to heuristic approaches such as knowledge-based expert systems. Practical real-world implementations of specific procedures (e.g., Kalman filters, Bayesian inference, and Dempster–Shafer evidential reasoning) exist. Algorithm selection criteria and the requisite *a priori* data are still major challenges, however. Applying Bayesian inference, Dempster–Shafer theory, artificial neural networks, fuzzy logic, Kalman filtering, or other data fusion algorithms to vehicle and event detection, classification, identification, and state estimation requires expert knowledge, probabilities, or other information from the analyst or data fusion specialist as discussed in Sections 2.7 and 2.8 as well as in subsequent chapters.

2.10 Level 2 and Level 3 Fusion Processing

This section provides illustrations of how the Level 2 and Level 3 processing concepts originally developed from a military perspective can be adapted to traffic management and ITS.

2.10.1 Level 2 processing

Level 2 processing (situation assessment) adopts a multiperspective point of view to identify the probable situation causing the observed data and events by combining the results of the Level 1 processing with information from other sources and databases. For traffic management purposes, these resources may include highway patrol or local police reports and databases, roadway configuration drawings, local and national weather forecasts, anticipated traffic mix, time-of-day and seasonal traffic patterns, construction schedules, and special event schedules.

In terms of a traffic management application, Level 2 fusion includes:

- Object aggregation—temporal, geometric, communications, and functional dependence relations among pedestrians, private and commercial vehicles, transit riders, bicyclists, work zone personnel, first responders, and roadside objects.
- Event and activity aggregation—temporal relations among drivers, other travelers, private and commercial vehicles, work zone personnel, road detours, lane closures, fleet operators, first responders, and other entities to identify meaningful events.
- Contextual interpretation—data analysis with respect to the context of the evolving situation, including weather, ambient light conditions, terrain, road and bike lane configurations, road topography, local driving habits, local traffic management strategies, availability of first responders, line-of-sight restrictions, time-of-day and seasonal traffic patterns and vehicle mix, special event occurrences such as sports and concerts, etc. Contextual analysis requires large databases. This, in turn, mandates the need for a balance between fast data insertion and fast data retrieval. Placing a time tag on database entries aids in determining the relevance of inferences drawn from the database.

2.10.2 Level 3 processing

When applied to traffic management, Level 3 processing (impact assessment) develops an impact-oriented perspective to estimate vehicle and driver capabilities, incident opportunities and timing, vehicle and driver (also pedestrian and bicycle as applicable) intent, and danger levels. It assesses the traffic flow patterns and other data to determine the likelihood and time of occurrence of a traffic event (e.g., traffic congestion, incident, construction or other pre-planned special events, and fire or police action) that impacts traffic flow. Information is analyzed to categorize the event as one that is imminent (i.e., seconds away), one that may occur within several minutes, or one that may occur when a longer time period has passed. Potential actions and their impacts on the event are evaluated.

As it pertains to traffic management, Level 3 fusion performs:

- Capability estimation—prediction of size, location, and capabilities of private and commercial vehicles, drivers, other travelers, and first responders; and the presence or lack of fixed roadside objects (e.g., trees, sign structures, luminaires, and guard rails) and randomly appearing objects (e.g., debris, animals, jay walkers, and bicyclists not in prescribed lanes).

- Prediction of vehicle and traveler intent—based on actions, driving culture, road configuration (e.g., presence of medians, multiple lanes, roundabouts, hills, curves, and elevation changes), entrance and exit ramp locations, congestion level, location of crosswalks and bicycle lanes; behavior patterns, such as the tendency of bicycle riders to travel on sidewalks; traffic signal timing; pre-planned special events, etc.
- Incident identification—based on driver and other traveler actions; appearance of first responders; and weather, data, and imagery from roadside sensors and other sources.
- Multiperspective assessment—data analysis with respect to first responders, commercial vehicle fleet operators, transit system operators, traffic management agencies, travelers, and other perspectives.
- Offensive and defensive analysis—prediction of results of hypothesized scenarios involving private and commercial vehicle drivers, other travelers, first responders, levels of congestion, and signal timing modifications.

2.10.3 Situation awareness

Lambert[91] proposes a seven–building-block approach for developing machine-based situation and impact assessment systems. One of these blocks, the symbolic imperative, introduces the interesting notion of *semantic registration* as the higher-level counterpart to spatiotemporal registration of data used in Level 1 processing. He states that "semantic registration involves the translation of object assessments and higher-level assessments into a canonical language accessible to higher-level fusion. But which symbols are required for this canonical language?" Lambert's suggested language for military-related, high-level fusion systems draws from the five groups of descriptors in Table 2.4, shown with examples of each in the left and middle columns. The right column contains examples of the descriptors that conceivably could apply to traffic management applications.

Other research concerning situation assessment examines the broader subject of situation awareness (SAW) and its decomposition into situation assessment and impact and threat assessment. SAW is described by Endsley as a state "in the mind of a human" that concerns "the perception of the elements in the environment within a volume of time and space, the comprehension of their meaning, and the projection of their status in the near future."[92,93]

A related concept is situation analysis, which Bosse, Jousselme, and Maupin[93] define as a process that examines "a situation, its elements, and their relations to provide and maintain a product, i.e., a state of situation awareness

Table 2.4 Descriptor groups and examples of semantic registration terms for machine-based high-level fusion.

Descriptors	Military Application	Traffic Management Application
Social	group, ally, enemy, neutral, owns, possesses, invites, offers, accepts, authorizes, allows	platoon, ally, wreckless, dangerous, neutral, owns, possesses, invites, offers, accepts, authorizes, allows
Cognitive	individual, routine, learnt, achieves, performs, succeeds, fails, intends, desires, believes, expects, anticipates, senses, informs, effects, approves, disapproves, prefers	individual, routine, learnt, succeeds, fails, intends, anticipates, senses, informs, effects, approves, disapproves
Functional	senses, moves, strikes, attaches, informs, operational, disrupted, neutralized, destroyed	senses, moves, informs, operational, disrupted, damaged, nonfunctional
Environmental	land, sea, air, outer space, cyberspace, incline, decline, number, temperature, weight, energy	interstate, urban, rural, paved, unimproved, multilane, divided, bridge, tunnel, downhill, uphill, straight, curve, number, temperature, visibility, weight
Metaphysical	exists, fragment, identical, time, before, space, connect, distance, area, volume, angle	exists, identical, time, in front of, behind, to the left, to the right, connect, distance, area, angle

for the decision maker." Roy offers another definition of situation analysis as "a process, the examination of a situation, its elements, and their relations, to provide and maintain a product, i.e., a state of SAW, for the decision maker."[41,94] His comprehensive model captures the relationships among the elements that comprise the situation and thus lays the foundation for comprehending the situation. The model includes situation element acquisition, data alignment and association, situation element perception refinement, situation element contextual analysis, situation element interpretation, situation classification and recognition, situation assessment, situation element projection, impact assessment, situation watch (observation), and process refinement.

Blasch, Kadar, Salerno, et al.[95] review several SAW models, including the situational awareness model (SAWM), user fusion model (UFM), perceptual reasoning model (PRM), and a combined situational awareness model and JDL model (CSAWM-JDL).

- SAWM provides situation awareness that assists a human-in-the-loop to determine his actions. Situational awareness is defined according to Endsley's definition[92] and has three levels: perception of elements in the environment, comprehension of the current situation, and projection of future states.
- The UFM allows a client to access all of the processing levels in the JDL model as needed, for example, to translate data into knowledge or, at a minimum, into information that can be exploited to reason and

make decisions. The UFM supports analysis or assessment by the user in three ways:[31,95] (1) physical tasks such as inputting *a priori* estimates, querying for results, and manipulating the data; (2) cognitive mental operations to infer meaning, utilize experiences, and modify results based on context; and (3) information dissemination and communication of data through reports and analysis.

Roy describes another form of UFM that integrates data and information fusion along with knowledge-based system concepts in support of situation analysis.[76,96] In this model, depicted in Fig. 2.18, situation analysis is the process that leads to situation awareness, which in turn provides knowledge and perspective about the scenario of interest. Situation awareness consequently requires knowledge, data, and information. Knowledge leads to a need for knowledge engineering and its constituent components of data acquisition, representation, and validation as inputs to a knowledge-based inference system. Data and information serve as inputs to a data and information fusion system and, along with the knowledge-based system, create a situation analysis node. Situation analysis also requires reasoning and inferencing. Not all of the situation elements of interest to a decision maker may be directly observable from the available data. This is especially true of highly abstract types of situation elements, such as intent and the relationships between situation elements. Therefore, those elements that cannot be directly observed must be inferred, i.e., derived as a

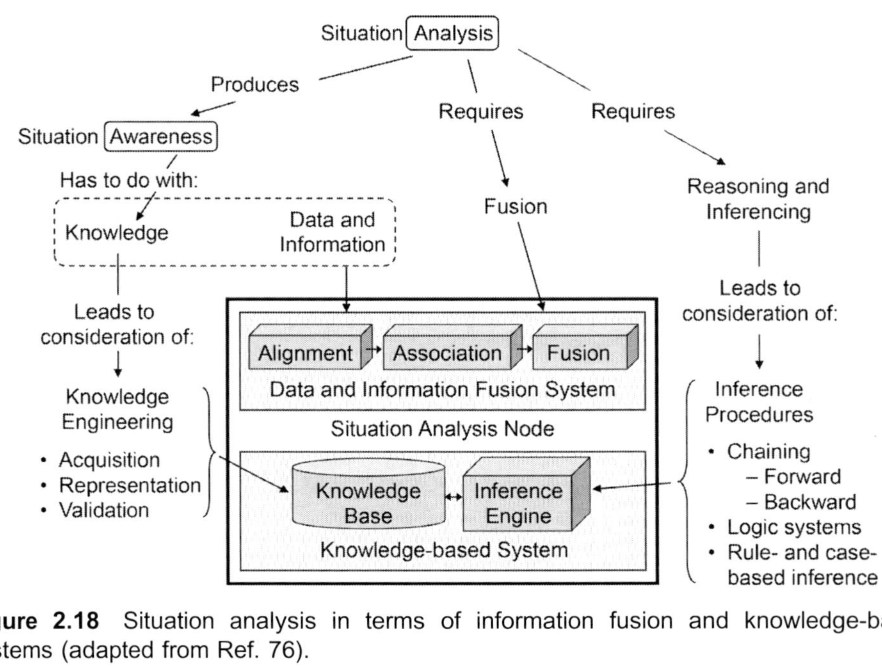

Figure 2.18 Situation analysis in terms of information fusion and knowledge-based systems (adapted from Ref. 76).

conclusion from facts or premises, or by reasoning from evidence. Accordingly, reasoning and inference procedures such as chaining, logic systems, and rule- and case-based inference become part of the knowledge-based inference system.

- The PRM is based on spatial and temporal associations. Its feedback planning and resource control system gather and assess current data and information, anticipate future hypotheses and beliefs, predict information requirements, and monitor intent. PRMs have the ability to allocate information and sensor resources that acquire data by controlling a resource manager function. Furthermore, a PRM is able to understand and act on acquired data by interpreting conflicting and misleading information in terms of the situation at hand.
- The CSAWM-JDL model allows a user to edit an existing template or model or to create a new one, activate or deactivate existing models, or view active models and any evidence that has been associated with the model over time.

2.10.4 Application to connected and self-driving vehicles

Data fusion processing Levels 2 and 3 play an important role in enhancing the safety and operability of connected and self-driving (SAE International's Level 4 and Level 5 automation) vehicles. Since these vehicles operate in a data-rich environment, they face the same challenges that all data fusion systems encounter, namely, information management and information fusion. Golestan, Soua, Karray, and Kamel[97] propose using information gathered from connected vehicles via vehicular *ad hoc* networks (VANETs) to identify vehicles and their inter-relationships. These authors observe that the convergence of VANETs and the Internet of Things has led to a new paradigm that they refer to as the Internet of Cars. VANETs permit vehicles to act as mobile sensor platforms that collect data from their sensed surroundings and then transmit relevant information to interested parties, e.g., drivers, pedestrians, bicyclists, traffic and transportation management agencies, first responders, and transit agencies.

Connected and self-driving vehicles must efficiently manage and interpret large quantities of diverse data and information to achieve SAW that alerts drivers to critical states, such as changing traffic flow conditions, erratic driving behavior, appearance of unexpected objects (such as jay walkers and animals), closed roads, and accidents. In viewing situation and impact assessment as part of the larger SAW paradigm, Golestan, Soua, Karray, and Kamel propose a SAW model whose top-level flow diagram is shown in Fig. 2.19 with the following descriptions:[97]

- Perception is characterized as an aggregation of entities and the relationships between them. Entities are identified in low-level data

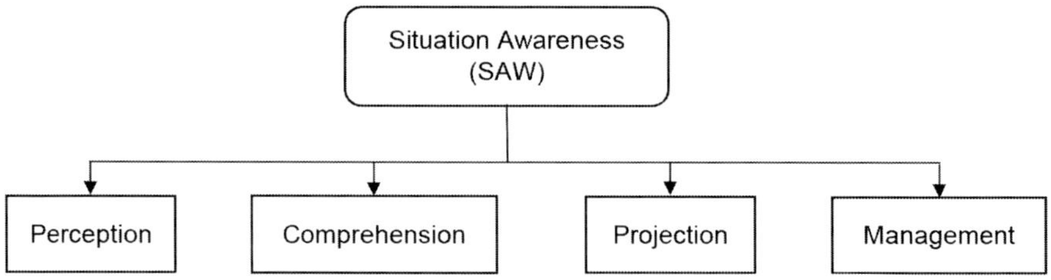

Figure 2.19 Situation awareness model of Golestan, Soua, Karray, and Kamel.

fusion (Levels 0 and 1) and include vehicles (single, platoons, moving, stopped, changing lanes), pedestrians, cyclists, and animals.

- Comprehension contains the situation and threat assessment modules. Situation assessment is the primary process that guides situation awareness. Threat assessment concerns measuring the capability, opportunity, and intent of a considered situation and, accordingly, serves as a bridge between situation assessment and impact assessment. The output of a threat assessment model is a set of hypothesized situations that are considered threatening and, therefore, can be used to assess the impact of a specific situation.
- The projection module contains the impact assessment and decision functions. Impact assessment acquires the previously assessed situations and their threats (i.e., the capabilities and intents associated with each situation) and calculates the likelihood of future hypothesized situations. A well-designed decision-making model should efficiently execute known *a priori* actions, e.g., steering and braking, given a set of hypothesized situations. Probabilistic graphical models, game theory, and machine learning methods such as decision trees are regarded as the main categories of decision-making models for impact assessment.
- Management's function is to facilitate collaboration between a SAW model and the human operator while refining the model's performance.

In related research, Golestan, Soua, Karray, and Kamel[98] and Golestan, Khaleghi, Karray, and Kamel[99] applied high-level fusion processing to analyzing information from connected vehicles via VANETs and used this concept to develop a collision warning system. Situation assessment provided the high-level inference relations and identified meaningful events, while impact assessment estimated future states and deduced the threat level (in traffic management parlance, the potential for an incident).

Since knowledge in these applications is complex, uncertain, and dynamic, questions of uncertainty management and semantics analysis arose. Therefore, the authors proposed fuzzy multi-entity Bayesian networks (MEBNs) to model knowledge for situation assessment, where first-order fuzzy logic defined the contextual and semantic constraints, and Bayesian networks processed data uncertainties and their dynamics.

Situations were defined as a set of entities that can be related in different ways and that are allowed to evolve temporally and laterally. The authors envisioned an impact model that contains three steps:

1. Situation *generation* to generate new situations conditioned on the current situation.
2. Situation *evaluation* as a measure for assessing and ordering the generated situations.
3. Situation *selection* to choose a subset of generated situations for a given impact assessment task.

In simulations of a collision warning system application, the authors were able to demonstrate the relationships between situations and threats, and how lateral and temporal behavior patterns assist in assessing events and activities.

2.11 Level 4 Fusion Processing

Level 4 processing (process refinement) seeks to enhance the effectiveness of the entire data fusion process by continuously monitoring and evaluating the ongoing fusion process using measures of performance and assessment. The Level 4 process also considers the need to relocate or add resources such as sensors and other data acquisition and analysis assets. For example, to aid in analyzing the information in cellular telephone calls and other voice communications, it may be necessary to use processes or algorithms that exploit syntactic and semantic formats and ontologies.[44] The data fusion community sees Level 4 processing as a meta-process, i.e., a process that monitors and seeks to optimize the overall data fusion process within operational and physical constraints.[43]

This is mirrored in the interpretation of Level 4 processing as a resource management function that Steinberg defines as a process "that combines multiple available actions, e.g., allocation of multiple available resources, over time to maximize some objective function."[100] A resource management process involves planning and control, and is one that:[9,100–102]

- Develops candidate response plans to respond to estimated world states (scenarios);
- Estimates the effects of candidate actions on mission objectives;
- Identifies conflicts for resource usage or detrimental side effects of candidate actions; and
- Resolves conflicts to assemble composite resource assignment plans based on the estimated net impact on mission attainment.

The duality between data fusion and resource management levels, displayed in Table 2.5, was proposed to improve the understanding of resource management alternatives and to provide insight into techniques that support the designs for each. The levels are partitioned in accordance with the

Table 2.5 Duality of data fusion and resource management processing levels (adapted from Ref. 9).

Level	Data Fusion Description	Resource Management Description
0	*Data assessment* detects, estimates, or perceives specific signals and features from the data sources.	*Signal management* tasks or controls resource response actions in the form of signals and observables, e.g., selection of sensor technologies to be deployed.
1	*Object assessment* detects, estimates, or perceives continuous parametric (e.g., kinematics, signatures) and discrete (e.g., class, type, color) attributes of object states.	*Resource response management* tasks or controls continuous and discrete resource responses, e.g., sensor operating modes, relocation of sensors and other data sources, choice of communications media.
2	*Situation assessment* detects, estimates, or comprehends relationships (e.g., aggregation, casual, command and control, coordination) among object states.	*Resource relationship management* tasks or controls relationships (e.g., aggregation, coordination, conflict) among resource responses.
3	*Impact or threat assessment* predicts or estimates the impact of Level 0, 1, 2 signals, objects, or relationship states.	*Mission objective management* establishes or modifies the objectives of Level 0, 1, 2 actions, responses, or relationship states.
4	*Process refinement* estimates system measures of performance (MoP) and effectiveness (MoE), and adjusts system resources or operational modes to meet objectives.	*Design management* tasks or controls system engineering and operational configuration.

"significant differences in the types of data, resources, models, and inferencing necessary for each level."[9] The objectives of resource management are to plan responses to improve the confidence in mission success and in the performance of the system. Furthermore, Steinberg and Bowman observe that, as with data fusion levels, "resource management levels are not necessarily processed in order and any one can be processed on its own or in combination given the corresponding inputs."[9] To further the comparison between data fusion and resource management, these authors appreciate that a resource management node performs functions that directly correspond to those of a data fusion node, i.e., data alignment; data association as decomposed into hypothesis generation, hypothesis evaluation, and hypothesis selection; and entity state estimation and prediction.[101] In resource management nomenclature, these functions become:

- Problem alignment that normalizes performance metrics for the given problem, normalizes performance models of the available resources, and aligns control formats and spatiotemporal data;
- Planning as decomposed into subtasks of:
  - Plan generation, consisting of candidate partitioning of the problem into subproblems and candidate assignment of resources;
  - Plan evaluation that calculates the conditional net cost, i.e., Bayesian cost (this can be expressed as the expected cost of

the plan execution minus the probability of an outcome multiplied by the value of the outcome); and
— Plan selection that determines a decision strategy; and

- Plan tasking and control, also called plan execution, that generates the control commands to implement the selected resource allocation plan.

Taking the duality argument one step further, resource management can be viewed as a data-fusion, systems-engineering problem that incorporates the above planning functions. Still, the data-fusion, systems-engineering process is not trivial since difficulties exist in the following areas: (1) representing the uncertainty in observations and in models of the phenomena that generate observations; (2) combining noncommensurate information, e.g., the distinctive attributes in imagery, text, voice, and non-imaging sensor data; and (3) maintaining and manipulating the substantial number of alternatives available to associate and interpret large numbers of observations of multiple entities.[100,101]

Process refinement requires that criteria exist for evaluating the performance of the system of interest. Performance measures often must accommodate varied and possibly conflicting facets that may be difficult to capture in one comprehensive and unified measure. This subject is explored further in Section 6.2.4.

In terms of a traffic management application, Level 4 fusion embraces:

- Evaluation—assesses performance and effectiveness of the fusion process. Were some types of data more useful in arriving at the correct understanding of the situation giving rise to the data? Did the data fusion process provide guidance that helped resolve the situation? Did the data fusion process serve the needs of all stakeholders, e.g., traffic management personnel; private vehicle drivers; first responders; transit operators, drivers, and passengers; and freight operators and drivers?
- Fusion control—identifies changes to processing functions that will improve performance. Does the structure of the data fusion algorithm need to be modified by adding additional types of classifiers to assist in identifying the objects or events in question?
- Source requirements processing—determines source-specific data requirements to improve multilevel fusion products. Is higher resolution or higher spatial density of sensors required? Are different types of data needed from the sensors to make the data fusion algorithm more effective in correctly interpreting the situation that gives rise to the data? Are the resources to acquire, store, and analyze big data from crowdsourcing, connected vehicles, and connected infrastructure available and used effectively?
- Mission management—recommends allocation of resources to achieve overall mission goals. Resources include roadside sensors; floating cars;

vehicle location tools that use GPS radios; Bluetooth MAC address, toll tag, and trucking industry transponder readers; DSRC transmitters and receivers such as those found in connected vehicles; ALPRs; and fixed and movable changeable message signs.

2.12 Level 5 Fusion Processing

Level 5 processing (user refinement) addresses questions concerning access to data and information, display of information, and human interpretation of the results of the data fusion process as they affect human decision making and actions.[38,39] User refinement is subsumed as an element of knowledge management within resource management where, among other functions, it adaptively determines which users query information, which have access to information, and which data are retrieved and displayed to support cognitive decision making and actions.

Blasch and Plano[39] develop a Level 5 Fuse-Act Situational User Refinement (FASUR) model that parses potential human behavior patterns that result from access to fused data into four categories: monitor, ignorant user, non-user, and effective user.

- The *monitor* is aware of what is occurring and of problems, needs, related factors, interactions, and data conflicts. However, this type of user is not able to act on the data and thus is impotent to do anything about what they see.
- The *ignorant user* is limited in their knowledge of what is occurring and is constrained in making a decision. Their capacity to act is seldom held back by their lack of perception, and they will act, even when it means mistrusting the displayed data.
- The *non-user* is blind to what is occurring and also impotent to do anything. These users do not care what data and information are available to them from the fusion system since they are not forced to act.
- The *effective user* possesses an understanding of the system that allows them to both fuse and act on the data and information presented to them. This user will find a cause for the observed data or events, will task sensors and other sources of information when more data are needed, or will take actions that implement process or knowledge refinement.

The Blasch and Plano model[39] also defines five functions that are needed as part of Level 5 processing: planning, organizing, coordinating, directing, and controlling.

- *Planning* defines the project and data acquisition requirements for a particular problem. In terms of systems engineering, it involves concepts exploration and benefits analyses, and development of system requirements, project plans, the systems engineering plan, and the concept of operations.

- In terms of a traffic management application, *organizing* entails specifying the types of sensors and other sources of data and information that will be used and allocating the required resources to the project. When multiple departments or agencies are involved in supplying data, interjurisdictional memoranda of understanding may have to be developed and approved if they do not already exist.
- *Coordinating* actions begins with the execution phase. Coordinating secures and maintains the cooperation of agencies that supply sensors and other data sources within and outside the user's control. Coordinating also involves finding and securing the best combination of data sources available.
- *Directing* implements the agreed-upon operations plan that utilizes the sources of data and other resources to achieve the established objectives. Directing is a combination of activating, tasking, delegating, monitoring, and refining the data acquisition, analysis, and presentation processes as needed.
- *Controlling* provides the formal measurement, analysis, and refinement of user actions at established intervals. Controlling requires users to establish metrics, to ensure that the results of the fusion process are providing the information needed to perform the mission at hand, and to take corrective action when necessary.

Nilsson and Ziemke discuss the broader impacts of human–computer interactions in terms of cognitive science and information fusion systems.[40,103] Research into cognitive science focuses not only on a single individual's internal thought processes, but also on the interactions with the surroundings, including other individuals and groups, artifacts, and additional types of information systems. Accordingly, Nilsson and Ziemke consider cognition as distributed in a three-fold sense: (1) across individuals in a group or organization; (2) between internal mechanisms in the human organism (e.g., memory and perception) and external mechanisms that include computer systems, material, and social and cultural environments; and (3) over time.

Human–computer interface (HCI) functions provide the mechanisms through which the results of fusion processing are conveyed to one or more human operators or analysts, and the means whereby an operator controls and guides the fusion inference process. Data must be accessible to a user or to multiple users in a timely fashion without overwhelming the user with constant interruptions from incoming data or extraneous information.

Fundamental design questions for HCI systems are: What does the user need to know, and when does it need to be known? Another complicating factor is due to the magnitude and variety of data that can be displayed, including fixed and free-text message formats under multiple protocols, and asynchronous, out-of-sequence, and false sensor reports.

Operators of data fusion systems should also be aware of human performance costs linked to particular forms of automation. The costs are reduced situation awareness, complacency, and skill degradation.[104] The first of these occurs when automated decision making reduces the operator's awareness of the system and of specific dynamic features of the work environment. If a decision aid, expert system, or other type of decision automation repeatedly selects and executes decision choices in a dynamic environment, human operators may not be able to sustain an acceptable understanding of the information sources because they are not actively engaged in evaluating the sources leading to a decision. The second cost arises if automation is highly, but not perfectly, reliable in executing decisions. This may cause operators to be complacent in monitoring the automation and its information sources and, hence, bring about a failure to detect the occasional times when the automation fails. Complacency is greatest when the operator is engaged in multiple tasks. Complacency also occurs if the algorithms used for data and information analysis are reliable, but not perfectly so. The third detriment appears when the decision-making function is consistently executed by automation, leading to a time when the human operator will not be as skilled in performing that function.

Other challenging issues arise concerning HCI design for traffic management applications. Since these fusion systems may operate in a stressful environment where people's lives may be at stake, such as when a traffic accident occurs, they should guide traffic management personnel through an effective decision-making paradigm in the face of stress, i.e., serve as a decision support system (DSS). Where several agencies are involved in resolving an incident or managing traffic flow, shared situation awareness is vital, and it is necessary to achieve a common state of understanding within all agencies through the exchange of data and information.[105] This requires that the lead agency's intent be accessible and understandable, and that the parties involved understand that shared situation awareness can only be developed over time and with practice.[40] There are also distinctive decision-making styles employed by different agencies that affect the way the agencies search for relevant data and information, and the way they perform analysis procedures.

The use of an information fusion system as a DSS involves the fusing of data and information from databases, models, and simulations.[104,106] The benefits of this approach are the following:

- The DSS can be developed to support different tasks, e.g., its function can be specific or general purpose.
- Both data-driven and user-driven actions can be supported. Data-driven interactions occur when data are received from sensors and are

processed by automated fusion processes for presentation to the user. User-driven actions occur when the user controls the system and initiates requests to retrieve information, perform computations, control the system resources, and provide input to assist the system.

- Information fusion systems can accommodate dynamic environments characterized by real-time data, ill-structured problems, shifting goals, and multiple users, in contrast to traditional DSSs that view humans as rational beings capable of making objective decisions among alternatives.

Notwithstanding the above benefits, challenges remain in treating DSS from an information fusion perspective. Nilsson and Ziemke enumerate these challenges as follows:[106]

- There is a need to possess a thorough understanding of the decision-making process to build an effective DSS. However, information fusion research is lacking focus on the user's decision-making process as embedded in an information fusion setting.
- Independent development of the information fusion system and DSS disciplines have created a need for informing and combining the research results from the two areas. As a case in point, the information fusion community could be guided in how to provide a user perspective when designing information fusion systems. Similarly, the DSS community could benefit from the information fusion community's viewpoint that an efficient information system that supports decision making needs the capability to present past, present, and future information (thus, integrating databases with simulations and models).

Table 2.6 summarizes HCI issues by listing the categories that influence user interactions, specific factors associated with each category, and the constraints often imposed when attempting to implement the functions performed by an information fusion system.[40] Constraints may affect multiple factors. A case in point is 'Provide access to original data and fused data,' which appears as a constraint in 'Decision making' and 'Multiple information sources' in the table. Table 2.6 also indicates the flow of information between categories. For example, the external environment, comprising sensors, databases, and the organization's functional relationships, all affect the users in terms of their cognitive abilities and the activities they can perform. The users' cognitive abilities, in turn, often limit the possible tasks they can execute. The trust factor relates to the acceptance level on the part of the user to the automated output of the particular tool. Management and users must recognize that trust evolves over time. The user exploits the interface to assist in completing various activities and, consequently, the interface is required to access the functions supported by the information fusion system. Lastly, the

Table 2.6 Human–computer interface issues in information fusion.

Category	Factor	Constraint
External environment	Organizational demands	Enable different levels of information availability to facilitate access for individuals and groups with different authorizations and job descriptions. Provide option to protect sensitive data. Capture organizational information that guides interaction to inform users. Encourage role-based systems. Integrate information fusion system into those currently operational within the organization.
affects	Multiple decision makers	Provide overlapping information to facilitate communication among team members. Use similar language to facilitate team communication. Introduce standard and advanced functions to meet varying user needs.
	Risk	Introduce thresholds to facilitate similar user decisions. Provide guidelines on how to respond to probabilities and other information provided.
	Temporal aspects	Clearly indicate temporal data, e.g., time and date, on displays to aid users.
	Dynamism	Provide flexibility in system for evolving requirements and tasks.
	Environment	Indicate if and how sensors are affected by environmental factors.
User's cognitive abilities	Cognitive issues	Allow interface personalization. Direct user's attention toward specific areas of interest. Restrict distracting clutter so as not to overload users. Focus on a subset of information to reduce cognitive workload. Support user's mental model for the system. Limit amount of data that needs to be processed simultaneously.
determines	Situation awareness	Provide alternative views of the situation at hand. Enable switching between a detailed or local view and a global view. Show your own situation in relation to that of others.
	Trust	Present uncertainty in information provided by automated process. Provide transparency to enable understanding of recommendations and predictions. Direct user training toward confidence building (in the outputs of the tool) rather than general training to encourage the buildup of user trust over time.

(*continued*)

Table 2.6 *(Continued)*

Category	Factor	Constraint
User activities	User tasks	Provide users interaction opportunities.
		Filter information but keep original data available for users who desire it.
utilize		Do not allow the design of the information fusion system to interfere with user tasks.
	Decision making	Provide a fit between decision makers and the decision-making process at the information fusion system output.
		Incorporate explanatory capabilities, feature matching strategies, and story generation or exploration according to the decision at hand.
		Enable filtering options to extract relevant information according to the decision at hand without hindering access to nonfiltered (original) data.
		Provide access to both fused data and original data.
		Facilitate fast decisions through easy access to certain information without a requirement for interaction.
Interface *access*	Input/output devices	Use multiple modalities to support simultaneous processing of information.
		Present data in visual form when possible.
	Visualization	Visualize uncertainty, information reliability, and quality of information.
		Display past, present, and future (predicted) information.
		Present different levels of abstraction or granularity in time and space.
Information fusion system *captures*	Multiple information sources	Indicate type of source when using multiple information sources to aid interpretations.
		Provide access to both fused data and original data.
	Uncertainty	Convey uncertainty (when it exists) in the information provided to others.
External environment and completes the cycle	Information flow	Provide flexibility to support both a top-down and bottom-up approach when required.
	Automation	Automate tasks that computers do best.

information fusion system itself captures various aspects of the environment in which the system operates.

2.13 Applications of Sensor and Data Fusion to ITS

Now that the reader is familiar with several data fusion algorithms and the data fusion process in general, we proceed to describe how these enhance the effectiveness of ITS and their subsystems. Many of the applications described below originally appeared in El Faouzi, Leung, and Kurian.[1] Other examples have been added to highlight the ongoing application of sensor and data fusion to ITS.

2.13.1 Advanced Transportation Management Systems

Advanced transportation management systems (ATMS) incorporate ITS technology in the planning, programming, and evaluation of transportation facilities to better enable them to respond to recurring and nonrecurring congestion and the demands of the traveling public. These facilities include: (1) freeways with ITS strategies such as ramp metering, information dissemination for travelers, managed lanes, and active transportation and demand management; (2) arterials using ITS strategies such as real-time traffic adaptive signal-timing plans, transit and emergency vehicle signal priority, and parking guidance; (3) integrated corridors; (4) road-weather systems; and (5) transportation management centers that may collocate with other organizations, e.g., local police, fire, rescue, and highway patrol agencies, in the same facility.[21,107]

The fuzzy ramp-metering algorithm of Taylor, Meldrum, and Jacobson[108] provides metering rate control based on local occupancy and speed data. The input variables utilized for the fuzzy control system are mainline and ramp occupancy, and mainline speed computed from the previous 20-s period data, in addition to ramp occupancy from additional past data samples. The output variable is the metering rate.

Seventeen production rules were heuristically developed based on experience in operating the ramp-metering system. A fuzzy centroid calculation converts the fuzzy values found from application of the production rules into a fixed and discrete output that can be used by the control system. The fuzzy ramp-metering algorithm was evaluated with FRESIM (FREway SIMulation) over a section of freeway that did not have many alternative diversion routes. The evaluation compared the fuzzy approach with other ramp meter controllers built into the FRESIM model, namely, clock metering, demand and capacity metering, and speed metering. The performance criteria were maintenance of a reasonable ramp queue, maximizing distance traveled, minimizing travel time (or equivalently maximizing average system speed), and minimizing the average delay and distance traveled per vehicle for all vehicles in the system. The more demand exceeded capacity the greater was the ability of the fuzzy controller to balance the conflicting demands of mainline efficiency and minimum ramp queues as compared to the other controllers. This was attributed to the ramp occupancy input data that was available to the fuzzy controller but not to the others. Generally, fuzzy control produced a modest improvement over the other metering controllers and a no-metering option.

Niittymaki and Kikuchi[109] designed a fuzzy system to control pedestrian crossing at a signalized intersection. The system mimics an experienced crossing guard who regulates the pedestrian movements. System objectives are to minimize pedestrian waiting time by accommodating the pedestrian as soon as possible, to minimize delay to vehicular movement by not stopping the vehicle flow for an unreasonably long period, and to maximize vehicle and pedestrian safety by preserving and allowing groups of approaching vehicles to pass. The input variables are as follows: cumulative waiting time of one or more pedestrians from the last signal change; the larger of the approach volumes (i.e., number of vehicles) measured by the two sensors used to detect approaching vehicles; and the smaller of the discharge gaps (i.e., time headway between two vehicles) measured by the two sensors in seconds. The output variables are the decision to extend the current signal phase or to terminate the current phase. A series of 18 production rules is used. Defuzzification is performed by selecting the decision corresponding to the output of the rule that contains the maximum value of the logical product.

In addition to the AID applications discussed in the next section, transportation management centers also apply data fusion to traffic state prediction and active traffic management (e.g., dynamic speed limits, hard shoulder running, and queue-end automatic detection).

2.13.2 Automatic incident detection

Incident detection algorithms for automatic recognition of incidents, accidents, and other road events requiring emergency responses have existed for more than three decades. Most of the algorithms were developed to operate with loop detector data. However, these algorithms exhibit mixed success. Interest in incident detection algorithms has been renewed partly because of the availability of new sensors and data sources, including probe and connected vehicles. Consequently, the availability of multisensor data puts AID in the class of problems that can benefit from data fusion methods.

Several data fusion techniques support incident detection and traffic management in general, including Dempster–Shafer evidential reasoning, Bayesian inference, and artificial neural networks. Most of the AID and traffic management applications combine probe vehicle data with conventional traffic sensor data. For example, Koppelman, Sethi, and Ivan,[110] Ivan, Schofer, Koppelman, and Massone,[111] and Ivan[112] developed an AID system using surveillance data from ILDs and probe vehicles that report link travel time. Two neural network strategies were investigated as potential data fusion procedures. The first strategy directly combined the observed traffic data from each source to determine if an incident was occurring. In the second strategy, the data from each source were pre-processed by separate incident detection algorithms, which reported scores that were then combined by the neural network. The application of neural networks to probe and detector-based incident detection on arterial networks offers promise for improved performance and reliability.[112] Dempster–Shafer evidential reasoning was also used to implement an operational AID system in studies by Byun, Choi, Ahn, and Ko.[113]

Thomas[114] investigated AID from a multiple-attribute, decision-making viewpoint with Bayesian scores. The author utilized combinations of probe travel times, number of probe reports, and ILD-reported occupancies and volumes as the inputs. The results show that models based solely on probe data lack in performance due to excessive overlaps in class distributions, while models based on detector occupancies and vehicle counts by lane perform well. It was also noted that the use of probe data enhanced the performance of sensor data-based models. Klein[115] and Klein, Yi, and Teng[116] studied the application of the Dempster–Shafer algorithm to a DSS that identified incidents and other events of concern to traffic managers. Additional information concerning this project is found in Section 4.8.

Increasing incident detection and classification accuracy from data fusion is still an area requiring improvement. Cohen's[117] investigations applied three aggregation schemes: logical aggregation, neural network fusion, and a veto procedure. The validation step, which was based on real-world data, demonstrated that both logical aggregation and veto procedure outperform the single best algorithm.

2.13.3 Network control

Data fusion techniques have been applied to construct an adaptive online traffic control method for urban and freeway road networks. Mueck's[118] model determines the queue length from vehicle counts produced by traffic flow sensors located close to the stop line and from signal timing information. Friedrich and Minciardi[119] introduce an approach based on queuing theory models for real-time queue length determination. In this method, Mueck's model serves as a quasi-measurement that utilizes Kalman filtering.

Wang and Papageorgiou[120] explored traffic state estimation on freeways using the EKF. Li, He, Yang, et al.[121] studied two strategies for traffic state prediction with multisource data and four machine-learning prediction models. Data consisted of microwave sensor time-mean speed and cellular probe space-mean speed. The first strategy combined the different types of data into a vector that served as the input to the prediction models. The second strategy applied the data from the two sources as separate inputs to the prediction models whose combined outputs give the final prediction value. The four prediction models were (1) a fully connected, feedforward ANN with one hidden layer; (2) support vector regression (SVR); (3) classification and regression tree (CART)—a binary decision-tree algorithm; and (4) k^{th} nearest neighbor (KNN). Li, He, Yang, et al.[121] conclude that the performance of the prediction models is different for the two strategies. Preliminary results indicate that the ANN and CART predict the general trend of speed but fail to capture the fluctuation in its value. Therefore, the ANN and CART are not optimal for short-term prediction, whereas SVR and KNN can predict both the trend and the fluctuation. Moreover, KNN is easier to implement and outperforms the other three models under both strategies as it can handle multisource data and provide accurate and current space-mean-speed prediction. Strategy 1 is preferable if ANN and SVR are used to predict the space-mean speed, while strategy 2 is preferable if CART and KNN are chosen.

Kong, Li, Chen, and Liu[122] assessed which of five different traffic states was operative in an urban setting using Dempster–Shafer evidence theory to combine estimates of mean speeds provided by single loops on a link of roadway (evidence set E_1) and mean speeds provided by 10,000 GPS-equipped taxicabs serving as vehicle probes (evidence set E_2). The loops were part of a SCATS (Sydney Coordinated Adaptive Traffic System) system in Shanghai, China. The five traffic states were: (1) *very congested* with speeds less than 1/6 $V_{\max}$, (2) *congested* with speeds between 1/6 and 1/3 $V_{\max}$, (3) *medium* with speeds between 1/3 and 1/2 $V_{\max}$, (4) *smooth* with speeds between 1/2 and 3/4 $V_{\max}$, and *very smooth* with speeds greater than 3/4 $V_{\max}$, where $V_{\max}$ is the speed limit assigned to the road. In their proposed algorithm,[122] probability masses were computed using a modification of a method proposed by Denoeux.[123] The Denoeux model expresses probability mass as a constant multiplied by an exponential function, where the exponent is proportional to the negative of the distance (based on some distance measure) between an input data sample and the kth nearest neighbors that belong to a specific class, in this case, a traffic state. This method is explained further in Section 4.7. The Kong et al. modification multiplied the probability mass by a dynamic reliability factor calculated using federated Kalman filters.[124,125] This multiplicative factor reduced the noise in the loop detector data, thereby smoothing the high-frequency fluctuations in the loop mean-speed data and enabling more suitable traffic state information to be presented to traffic managers and drivers of vehicles.

Gao, Ozbay, and Kurkcu[126] also applied the Denoeux method for finding probability mass functions to the fusion of traffic speed data collected by the New York City Midtown in Motion (MIM) adaptive traffic signal control system, toll tag readers, and GPS-equipped buses. Video camera imagery provided the ground truth data. The reliability factors in this study were based on the relative applicability of the data to individual blocks within a travel segment, typically eight blocks, and the sample size. The Dempster–Shafer approach was found to be more accurate than two other fusion algorithms, namely, a weighted method that assigns static weights to each of the data sources and a Random Forest, supervised nonlinear classification algorithm.

2.13.4 Advanced traveler information systems

Advanced traveler information systems (ATIS) employ a variety of automatic data collection techniques to assist in understanding traffic conditions and derive relevant indicators to support traveler guidance.[127] Traveler information is often presented as travel time and serves as a measure of impedance (or cost) for route choice strategies. Travel time is also used by network operators as an indicator of quality of service (QoS). However, estimating travel times with an acceptable degree of accuracy may be difficult in urban areas because of theoretical, technical, and methodological challenges encountered when using roadway-based sensors. For example, these sensors provide spatial sampling that is dependent on the sensor density and ultimately on cost considerations. Furthermore, such measurements only represent traffic flow conditions at the sensor location and not over the entire link. Consequently, roadway-based sensors that measure the prevailing traffic conditions on an urban road are frequently ineffective. The proliferation of alternative data and information source devices (e.g., surveillance cameras, GPS receivers, cell phone tracking, license-plate readers, probe vehicles, and connected vehicles) provides alternatives that supplement the data offered by conventional roadway sensors. These alternative sources also have the potential to improve the accuracy of travel-time estimates. As a result, travel-time estimation becomes a candidate data fusion problem.

Tarko and Rouphail,[128] Thakuriah and Sen,[129] and Berka, Tian, and Tarko,[130] discuss the requirements for data fusion in the ADVANCE program. ADVANCE was an in-vehicle application of ATIS that provided route guidance in real time in the northwestern portion of Chicago and its northwest suburbs. It used probe vehicles to generate dynamic travel-time information on expressways, arterials, and local streets. Data from loop detectors and the travel-time reports were combined using inference rules. Evaluations of the proposed algorithms found that probe data greatly improve static (archival average) link travel-time estimates by time of day. Dailey, Harn, and Linet[131] describe a data fusion algorithm for ITS that estimates speed from volume and occupancy measurements.

Abe[132] describes advances in travel-time forecasting using data from automatic vehicle identification devices to correct the forecast. Dynamic route guidance systems (DRGS) are another potential area for travel-time estimation and prediction through data fusion. Kühne[133] offers a framework for fusing information from various sources within DRGS. The data sources consist of loop detectors, probe vehicles, and a QoS indicator with some exogenous information concerning road works and incidents. The proposed solution is based on a weighted mean scheme. The weights are derived according to the source reliability. Choi[134] and Choi and Chung[135] address the problem of generating travel time from loop detectors, probe vehicles, and video camera sources using fuzzy logic.

El Faouzi and Lesort[136] and El Faouzi[137] propose an estimation framework for real-time traffic condition characterization based on multi-source data. As an illustrative example, multisource travel-time estimation was implemented using two data sources: conventional loop detectors that deliver Eulerian data (i.e., data at a fixed location) and probe vehicles that collect Lagrangian data (i.e., data from sources that move with the traffic flow). Travel-time measurements collected by license plate matching served as reference or ground-truth data for validation purposes. Loop detector data were statistical in nature and were analyzed from a distributed estimation problem point of view: each source derives an estimator of travel time, and the individual estimates are then combined using a weighted mean strategy. The weights were derived from variance–covariance estimation errors. Results show the need to improve the estimation accuracy of the proposed schemes.

In another approach, El Faouzi[138,139] and El Faouzi, Klein, and De Mouzon[140] describe the application of Dempster–Shafer evidential reasoning to calculate travel time on a roadway. The first two of these references use field test data from ILDs and probe vehicles on an arterial, and the third uses field test data from ILD pairs and electronic toll tags on vehicles over a segment of a controlled-access highway. Ground truth for the first project was travel time collected using license plate matching, while ground truth for the second came from analyzing all available toll collection data (i.e., a combination of electronic toll tag, real-time credit card, and cash payments). The probability masses for each data source (ILDs and probe vehicles in the arterial study and ILD pairs and electronic toll tags in the controlled-access highway study) were found by constructing confusion matrices formed by comparing the estimated travel times from the data sources with the ground-truth values, followed by a normalization procedure. Probability masses from the ILD and electronic toll tag confusion matrices were combined using Dempster's rule. The results displayed some improvement in travel-time prediction as compared with the values obtained from a single data source.

Dempster–Shafer reasoning was also employed by Zhang, Hamedi, and Haghani[141] to estimate travel time on a freeway using three types of data, namely, two Bluetooth MAC address readers, INRIX vehicle probes equipped with GPS location devices, and speed detection data from three microwave sensors maintained by HERE. These authors adopted a statistics-based approach to generate probability mass values for the travel-time intervals, assuming that the travel time for a given time interval follows a lognormal distribution. Due to the stochastic nature of traffic, significant fluctuations might exist during the data acquisition process. Therefore, to ensure input data quality, it was necessary to assess the reliability of the real-time data before calculating the probability masses. For travel-time data generated by identification and re-identification methods (e.g., Bluetooth MAC address readers in this case; however, electronic toll tag and license plate readers may also be used) and by extension to probe-based data, Zhang et al. utilized a scheme developed by Dion and Rakha.[142] The scheme employed a credibility factor α that was a function of the number of valid detections within the time interval of interest to quantify the weights given to the real-time and historical data eventually incorporated into the probability mass calculation for each travel-time interval. If the real-time data did not meet quality standards, then the historical travel-time distribution for the same time period was used to create a reliable estimate of the travel time. When travel-time data arose from a set of point detectors (e.g., microwave radar sensors), a different reliability function was needed due to the structure of the estimation method. In this circumstance, Zhang, Hamedi, and Haghani[141] assessed reliability with another exponential-smoothing factor related to the sensor spatial density and traffic flow fluctuations. This measure recognizes that travel-time estimation from point detectors is dependable only when the distance between two consecutive sensors is relatively short and the traffic flow is relatively stable. Their methods produced travel times from stationary detector estimation that were about 0.5 min higher during periods of noncongestion, and about 0.5 min lower during congestion. However, the Bluetooth and probe data were more likely to be equal during periods of congestion and noncongestion. In addition to the fused travel time, the model produced a numerical confidence score for each interval based on quantification of travel-time uncertainty for each data source appearing in the fusion scheme.

2.13.5 Advanced driver assistance systems

Passenger safety is a primary function of ITS. The increased availability of ADAS and collision avoidance systems (CAS) is indicative of the growth in active safety devices that complement traditional passive systems such as seat belts and air bags. CAS provide more-reliable data that are descriptive of traffic and other hazards surrounding the vehicle in pre-crash situations. The input data frequently come from a variety of sensors, including radar, lidar, and cameras. The fused data alert drivers to potentially dangerous situations and provide a static map of the environment and the vehicle position.[143]

Self-driving or autonomous vehicles are quickly moving from a research phase to an operational phase where data fusion is key to their operation. Automated and autonomous vehicles can potentially be deployed in hazardous and unknown environments and, of course, can be used for personal highway driving. In these applications, the vehicle senses its environment through an array of sensors whose outputs provide decision support assistance and, in the ultimate form, completely automated driving. Challenges include the heterogeneous nature of the data, extracting relevant features from the measurements that can be used in data fusion algorithms, and developing a robust set of algorithms and features that can function on a variety of roadway types with an assortment of traffic control devices and traffic flow conditions, and in inclement weather.

Murphy[144] discusses the role of sensor fusion in vehicle guidance and navigation, and proposes general methods for fusing data and sensor-fusion activities within a robot architecture. Pei and Liou[145] propose 3D vehicle motion estimation by fusion of multisource information. Image point and line features are also considered for fusion. Langheim and Buchanan[146] investigated data fusion systems for automatic cruise control with stop-and-go phenomena, while Stiller, Hipp, Rossig, and Ewald[147] report a data fusion framework for obstacle detection and tracking.

Tsogas, Dai, Thomaidis, et al.[148] applied Dempster–Shafer evidential reasoning to develop a driver assistance system that warns drivers if their vehicle changes lanes, cuts in front of another vehicle, is following a vehicle in the same lane, is following a vehicle in an adjacent lane, is changing lanes to overtake a vehicle, or is traveling in a free flow state. The system uses a Kalman filter to estimate the lateral velocity and acceleration of the ego vehicle (a vehicle equipped with an automated driving system), which are needed to calculate time and distance parameters that detect the maneuvers, namely, lane change, following vehicle in path, overtaking, following vehicle in next lane, cut in, and free flow. The Dempster–Shafer process determines the probability masses that correspond to each of the six maneuvers from data provided by a forward-looking camera; short-, medium-, and long-range radars; and inertial sensors mounted on the ego vehicle. The probability masses from each set of sensor modalities are combined by forming the orthogonal sum through the application of Dempster's rule as described in Section 4.5.1. When the orthogonal sum produces a null set, Tsogas et al. apply the Yager[149] technique for allocating probability mass from the conflicting propositions to the frame of discernment.

2.13.6 Crash analysis and prevention

Although there has been a steady reduction in the number of accidents, crashes continue to contribute heavily to losses in both human and economic terms. The reduced number of accidents could be due to multiple efforts: road infrastructure improvements, enforcing of more strict regulations on alcohol and speed, and improvements in vehicle design that contribute to vehicle safety. Sohn and Shin[150] exploited retrospective accident data, such as traffic accident records, as inputs to neural network and decision-tree algorithms to develop a classification model for road traffic accident severity (bodily injury or property damage) as a function of potentially related categorical factors. These authors noted that the classification accuracy of the individual algorithm was relatively low. Furthermore, Sohn and Lee[151] realized that the accuracy could be increased by applying data fusion and ensemble algorithms. Breiman[152–154] and Wolpert[155] explored the data ensemble method that combines various results obtained from a single classifier tailored using several bootstrap samples. Three approaches were investigated: (1) classifier fusion based on the Dempster–Shafer algorithm, Bayesian inference, and logistic models; (2) data ensemble fusion based on arcing and bagging; and (3) clustering based on the *k*-means algorithm. Empirical results indicate that a clustering-based classification algorithm works best for road traffic accident classification.

Oh, Oh, and Ritchie[62] studied a real-time accident warning system that incorporated Bayesian classification and a PNN to differentiate between normal and hazardous traffic conditions. The pdfs required by the Bayesian approach were estimated with nonparametric Parzen density functions. Input parameters to the PNN were the standard deviation of speed and average occupancy computed from speed and occupancy data collected every 10 s at double-loop detector stations. The authors' proposed methodology had the high identification rate (the percentage of total accidents identified) and low warning rate (percentage of time when the accident likelihood is above the given threshold) required of a practical warning system.

2.13.7 Traffic demand estimation

An important topic in the field of transportation planning and management is OD estimation from link counts. An estimation of a most likely OD matrix is typically derived from counts made on specific links of the transportation network. Some of the proposed schemes derive the OD matrices by combining data from different sources. An illustration of this class of problems is the dynamic OD estimation initiated by Cremer and Keller[156] and Cremer.[157] Further development along the same lines was pursued by Okutani[158] and Ashok and Ben-Akiva.[159] Kalman filtering is commonly applied in this class of problems. Ben-Akiva and Morikawa[160] explored OD estimation methods that combine different data sources (stated preference data and traffic measurements), and Lundgren, Peterson, and Tengroth[161] describe a method for adjusting time-dependent travel demand information by incorporating link flow observations. These authors utilized the structure of the given OD matrix, which was developed from different sources, to make simple overall adjustments.

Hyun, Tok, and Ritchie[162] applied a naïve Bayesian model to match individual freight vehicles traveling between two data collection locations. The technique has applications to freight modeling, pavement maintenance, vehicle tracking, and travel-time monitoring. Input data were truck physical attributes obtained from WIM sensors and waveform signatures produced by inductive loops sampled at up to 1200 samples/second. WIM attributes include gross vehicle weight, total vehicle length, time duration a vehicle occupies a loop, axle spacing, and axle weight. The high sampling frequency of the loops produces distinct waveforms from different types of metallic vehicles. However, several factors such as vehicle entrance angle, sensitivity of the detector, and geometry of the detection site cause variations in the signature. Since the attribute independence assumption is violated for attributes obtained from WIM sensors, attribute weights derived from entropy-based filters were used instead to preserve the independence assumption and to improve matching. Data for model development and the subsequent evaluation of its ability to track and re-identify trucks were obtained from different freeway segments. Model development data were collected from two WIM sites spanning 26 miles along Interstate-5 between San Onofre and Leucadia, California. Here only 11% of trucks observed at the downstream site traversed the whole corridor. The evaluation data came from two WIM sites separated by 65 miles along State Highway 99 in the San Joaquin Valley and had an 81% correct-matching rate for the trucks declared as through trucks by the algorithm. Up to 17 WIM classification attributes can be obtained from trucks due to differences in axle spacing and weight characteristics that vary by axle configuration. Inductive loop signature attributes include 50 normalized magnitude measurements obtained at evenly distributed points along the temporal axis of the inductive signature. In total, 67 possible features were available from WIM and loop signature data. Gaussian distributions were estimated for all 67 feature vectors. The loop signature feature vectors represent the differences between the signature attributes of a target vehicle (one detected at the downstream detection location) and a candidate vehicle (one detected at the upstream detection location). After a truck is detected at a downstream location, the tracking algorithm searches for candidate trucks that passed the upstream location at some earlier time. The Gaussian distributions for the feature vectors were exploited to distinguish matched and nonmatched pairs through a statistical test and a clustering method. The overall tracking accuracy, including both correctly matched vehicles and vehicles identified as nonmatches, was 91% in the evaluation test data set.

2.13.8 Traffic forecasting and traffic monitoring

Traffic flow forecasting is of importance to traffic surveillance and management. Many traffic flow prediction schemes were based on classic autoregressive models, especially time series techniques. Harrison and Stevens[163] explored this problem in the context of a Bayesian framework. Okutani and Stephanedes[164] used Kalman filtering techniques. Vythoulkas[165] used neural networks and system identification, and El Faouzi[166] used a nonparametric paradigm that incorporated kernel techniques. None of these approaches allows one to achieve highly accurate predictions except in some special situations (for some network configurations and/or networks with high sensor density). This is due to some extent to traffic dynamics that cannot be formalized by a single procedure. Therefore, one can obtain different forecasts of the same quantity from different methods (the important assumption here is that different predictors are measures of the same quantity or various aspects of the same item), which is a detriment for traffic operations where highly accurate forecasts are needed. The solution to this problem often takes the form of the single "best" predictor in some sense (most accurate values, most appropriate models of the underlying process, most cost-effective, etc.) among the available forecasting methods. Another tactic consists of combining the individual forecasts. Bates and Granger[167] and Granger[168] demonstrate that the linear combination of several predictors from a single data set can outperform the individual predictors. This conclusion is similar to that of Hernandez, Tok, and Ritchie[64] in their study of truck classification described later in this section.

El Faouzi[169] developed a way to combine various forecasts of the same quantity. He derived two predictors based on nonparametric traffic flow using a kernel estimator and predicting scheme founded on propagation of a lagged upstream traffic flow. Cremer and Schrieber[170] studied the integration of in-vehicle information and loop detector data using the EKF. Sau, El Faouzi, Ben Aïssa, and De Mouzon[171] investigated traffic monitoring and prediction with multisource data by applying a particle filter as the state estimation technique. Choi[172] examined the problem of missing data estimation and proposed an evidential reasoning framework for inferring the missing data.

A fully connected feedforward network was used by Sekula, Marković, Vander Laan, and Sadabadi[173] to estimate historical hourly volume profiles from average annual daily traffic data collected in Maryland. Their model consisted of an input layer, three hidden layers with 300 neurons each, and an output layer. The network was trained and evaluated with aggregate hourly vehicle count data collected from automatic traffic recording stations located along Interstates, U.S. routes, and state routes. Speed data from vehicle probes equipped with GPS radios, hourly weather data, infrastructure characteristics (number of lanes, speed limit, class and type of road), temporal information (day of week, time of day, holiday), and historical hourly volume profile data contributed additional features to the neural network. The probe data were particularly effective in improving the performance of the ANN.

Hernandez, Tok, and Ritchie[64] explored the integration of WIM system classification attributes and high-resolution waveform signatures (related to the undercarriage configuration) from ILDs to classify trucks into 14 classes based on axle count, axle spacing, and gross vehicle weight according to the

Federal Highway Administration (FHWA) classification scheme. This approach has the potential to improve freight movement forecasts, estimate truck payloads by body type, monitor air quality, and assist transportation agencies in managing infrastructure assets. A multiple-classifier system (MCS) combined the outputs of naïve Bayes, decision tree, support vector machine, multilayer feedforward neural network, and PNN classifiers to leverage the statistical and mathematical advantages of each classifier, while eliminating the constraints associated with single classifier use. The individual classifier predictions were combined with Bayes' theorem, assuming that the classifiers exhibit conditional independence given a class label.[174] The likelihood functions required for the Bayes process were obtained from a cross-classification matrix, i.e., a confusion matrix, created for each classifier. The class that gave the largest posterior probability was selected as the final prediction of the ensemble of classifiers. Using the MCS approach, each of the ten FHWA classes (Classes 4–12 and 14) achieved an overall correct classification rate above 80%. These results represented a significant improvement over the performance of the best individual classifier.

2.13.9 Position and heading estimation

Modern transportation systems require accurate information concerning the position and heading of the vehicle. Inertial navigation systems (INSs) that determine vehicle location rely on dead reckoning. The difficulty with these systems is the integration-induced accumulation of small errors in the measurement of acceleration and angular velocity into progressively larger errors in the position estimate. In the last few decades, the GPS, initially developed as a military navigation aid, has gained wide acceptance in civilian navigation systems.[175] When satellite signals are blocked by tall buildings or degraded by electromagnetic interference or ionospheric-caused diffraction, GPS outage or large errors occur. In such situations, the estimation of position is at best inaccurate and at worst impossible, and the device ceases to operate satisfactorily. Data fusion can effectively combat the drawbacks that occur when INS and GPS devices are used by themselves rather than in combination. The benefits of a combined GPS–INS device accrue from being able to calibrate the INS from GPS signals when they are available and using the INS to obtain position and angle updates at a quicker rate when the GPS is blocked or degraded.

2.13.9.1 GPS–INS applications to ITS

GPS–INS integration has benefited from the use of the Kalman filter since the 1990s. Wei and Schwartz[176] utilized a decentralized filtering strategy by applying a Kalman smoother to integrate the GPS differential range and phase measurements with INS data. Variants of the Kalman filter are often employed to improve integration. A constrained UKF algorithm was proposed by Li and Leung[177] to fuse differential GPS, INS (gyro and accelerometer), and digital maps to localize vehicles for ITS applications. Kalman filter operation requires accurate stochastic models for sensor measurement data and kinematic models of the vehicle's motion. Such requirements are often difficult to satisfy and result in the application of artificial intelligence techniques for GPS–INS integration. For example, multilayer perceptron and radial basis function neural networks were successfully applied for GPS–INS integration by El-Sheimy, Chiang, and Noureldin[178] and Sharaf, Noureldin, Osman, and El-Sheimy.[179] GPS–INS integration was also obtained with adaptive neuro-fuzzy techniques in Hiliuta, Landry, and Gagnon.[180] Such systems mimic vehicle dynamics by training the neural network during periods when GPS signals are available.

Toledo-Moreo and Zamora-Izquierdo[181] developed an interacting multiple model (IMM) with EKFs for change lane (CL) and keep lane (KL) maneuver states to provide vehicle lane positioning and maneuver predictions to traffic managers. GPS and INS data are inserted into the CL and KL models for two types of roads, one with negligible curvature and one that accounts for road curvature. In the latter model, the horizontal road alignment is described by approximations to clothoid functions whose curvature changes linearly with the traveled distance. Three-dimensional road aspects are not modeled based on the assumption that any highway vertical motions are negligible compared with those in the horizontal plane.

In the CL model for roads with negligible curvature, the state vector was represented by components consisting of east, north, velocity angle, velocity, yaw rate of turn, and acceleration in the center of mass of the vehicle. GPS data provide the east and north values, INS measurements provide the yaw rate and longitudinal acceleration, and odometry provides the velocity. The state vector for the KL model is the same except that a constant yaw is assumed between sampling periods such that the yaw rate of turn is zero. In the CL model that includes road curvature, the state vector adds two parameters to the CL state vector for the negligible curvature road, namely, white noise terms that represent the errors due to model assumptions of the road shape. The state vector of the KL model that includes road curvature is the same as in the CL model. However, the derivative of the velocity angle is assumed to follow the road shape such that the yaw rate of turn is proportional to the vehicle velocity.

Among their stated conclusions is the following: Maneuver state changes are typically predicted by the proposed IMM method in 0.2 to 0.3 s, which is about 5 times faster than using differential GPS and map-matching algorithms as reported in another study.[182] Where the road curvature was null or very small in the Toledo-Moreo and Zamora-Izquierdo study, the negligible road curvature CL and KL models were appropriate. However, their poor results on more curved trajectories urge the use of road shape data in the prediction process.

Accordingly, the vehicle and the road trajectory estimates incorporated into the second road model set achieved good results not only in straight road stretches, but under challenging conditions of different road curvatures and GPS outages. Additional information concerning IMMs is found in Section 5.12.

2.13.9.2 GPS modernization program

The future appears promising for the exploitation of GPS devices in monitoring and predicting vehicle position and heading. The U.S. GPS modernization program is adding new civilian signals and frequencies to GPS satellites, enabling ionospheric correction for all users and providing the technology needed to eliminate the accuracy difference between military and civilian systems.[183] A summary of the characteristics of the four GPS signals currently available or planned is found in Table 2.7. Dual-frequency receivers that receive the original L1 C/A 1575-MHz frequency and the more-recent L2C 1227-MHz frequency deliver faster signal acquisition, enhanced reliability, and greater operating range. L2C signals became available in 2005 and as of May 2018 are broadcasting from 19 GPS satellites. The signal should be accessible on 24 GPS satellites around 2021. These transmissions broadcast at a higher effective power than the legacy L1 C/A signal, making them easier to receive under trees and indoors.

Table 2.8 lists the data fusion algorithms currently embedded in ITS applications.[1,3,4,21] These include Bayesian inference, Dempster–Shafer evidential theory and some of its modifications, artificial neural networks, fuzzy logic, knowledge-based expert systems, and vehicle and pedestrian tracking based on the Kalman filter, EKF, UKF, and Monte Carlo techniques including the particle filter.

Table 2.7 U.S. GPS modernization program.

Frequency Symbol	Frequency (MHz)	Available	In all 24 satellites	Benefits
L1 C/A (coarse/acquisition)	1575	At present	At present	Legacy GPS signal.
L2C (Civilian)	1227	2005	2021*	Provide reception under trees and indoors.
L5	1176	2010	2024*	Used by aircraft with L1 C/A to improve ionospheric correction accuracy and signal redundancy.
L1C	1575	2018	Late 2020s*	Enable interoperability between GPS and international navigation satellite systems (European Galileo, Japanese QZSS, Chinese BeiDou) and improve mobile GPS reception in cities and other challenging environments.

*The dates are from the 2017 Federal Radionavigation Plan. At the time of this book's publication, the 2019 version of the plan was in final coordination for signature by the Secretaries of Defense, Transportation, and Homeland Security, and will likely include new dates.[184]

Table 2.8 Data fusion algorithms and architectures currently found in ITS applications.

Application	Data Fusion Algorithm	Architecture
Ramp metering	Fuzzy logic	Sensor level[1]
Pedestrian crossing time	Fuzzy logic	Central level[2]
Automatic incident detection	Artificial neural network	Sensor level
Automatic incident detection	Bayesian inference	Sensor level
Automatic incident detection	Dempster–Shafer	Sensor level or decision level[1]
Travel-time estimation	Inference rules	Sensor level
Travel-time estimation	Dempster–Shafer	Sensor level
Travel-time estimation	Weighted mean of several travel-time estimators. Weights are a function of the variance or covariance of the estimators.	Sensor level
Travel-time estimation	Weighted mean where weights are a function of the data source reliability.	Sensor level
Travel-time estimation	Fuzzy logic	Sensor level
Vehicle and object tracking	Kalman filter	Central level
Lane departure warning	Image processing using edge detection and extraction of other features.	Pixel level[3]
Traffic state estimation	Extended Kalman filter	Central level
Crash analysis and prevention	*k*-means algorithm	Sensor level or decision level
Traffic forecasting and monitoring	Bayesian inference	Sensor level
Traffic forecasting and monitoring	Artificial neural network	Sensor level
Traffic forecasting and monitoring	Kalman filter	Central level
Traffic forecasting and monitoring	Extended Kalman filter	Central level
Traffic forecasting and monitoring	Kernel estimator	Central level
Traffic forecasting and monitoring	Particle filter	Central level
Vehicle position estimation	Unscented Kalman filter	Central level
Vehicle position estimation	Artificial neural network	Central level

1. In sensor-level and decision-level fusion, each sensor detects, classifies, identifies, and estimates the state of the objects of interest before data entry into the fusion processor, which combines the individual sensor data to improve the classification, identification, or state estimate of the objects. Sensor- and decision-level fusion are optimal for detecting and classifying objects only if the sensors rely on independent signature generation phenomena to develop information about the identity of objects in their field of regard; that is, they derive object signatures from different physical processes and generally do not report a false alarm on the same artifacts.
2. In central-level fusion (also called centralized fusion), minimally processed sensor data are sent to a fusion processor that analyzes the data for object features or attributes that aid in identifying and tracking the objects.
3. In pixel-level fusion, minimally processed data from each sensor are combined at the pixel- or resolution-cell level of the sensors using a central-level fusion architecture. Little, if any, preprocessing of data occurs before the data are fused.

2.14 Summary

Sensor fusion is a term that describes a sensor or data source architecture utilizing multiple sensors to improve data accuracy or reliability, to extend the surveillance area, or to gather additional information about events or objects that are present in the area of interest. According to the JDL definition, data fusion is a multilevel, multifaceted process that addresses the automatic detection, association, correlation, estimation, and combination of data and information from single and multiple sources to achieve refined position and identity estimates, and complete and timely assessments of situations and their impacts and significance.

The combined JDL–DFIG data fusion model references seven levels of processing. Level 0 preprocesses data with techniques such as normalizing, formatting, ordering, batching, and compressing. Level 1 data processing identifies the objects or events and their state by combining or fusing data from all available sources. Level 2 processing identifies the probable situation causing the observed data and events by combining the results of the Level 1 processing with information from other sources and databases such as police reports, roadway configuration drawings, local and national weather forecasts, anticipated traffic mix, time-of-day and seasonal traffic patterns, construction schedules, and special event schedules. Level 3 processing assesses the traffic flow patterns and other data with respect to the likely impact of a traffic event on traffic flow (e.g., duration of traffic congestion, incident, fire, or police action). Level 4 processing seeks to improve the entire data fusion process by continuously refining predictions and performance assessments, evaluating the need for additional sources of information, and ensuring that the expectations of all stakeholders are met to the extent possible. Level 5 processing is concerned with enabling a human to control, interpret, and apply the results of the fusion process. Level 6 processing addresses mission management and involves adaptive spatiotemporal control of assets, route planning, and goal determination to support team decision making and actions that recognize social, economic, political, and organizational constraints. The taxonomy for object detection, classification, and identification algorithms includes physical models, feature-based inference techniques, and cognitive-based models. The taxonomy for state estimation algorithms identifies two general methods for searching and updating existing track files, namely, data-driven or track-driven, and identifies the techniques available to correlate and associate data and tracks.

The JDL data fusion levels are intended only as a convenient categorization of data fusion functions. Data fusion levels are not intended to be, nor should they be, executed in order of their levels. Processing should be partitioned in terms of the individual system requirements and to provide outputs that satisfy the needs of the decision maker.

Data fusion applications to transportation management have been ongoing for at least two decades. Level 1 data fusion algorithms that have been successfully applied to traffic management include fuzzy logic, artificial neural networks, Bayesian inference, Dempster–Shafer evidential reasoning, inference rules, weighted mean of several travel-time estimators, Kalman filtering and its variants, image processing using edge detection and extraction of other features, *k*-means algorithm, kernel estimator, and Monte Carlo techniques including the particle filter. Level 2 and 3 fusion applications are also appearing. For example, in England, the National Traffic Information Service is combining inductive loop speed and flow information with automatic license-plate reader travel-time data and accident, inclement weather, and work zone locations to generate predictive forecasts concerning the duration of an event.[185] The various situation awareness models developed for defense applications were described to encourage their exploration for relevance to traffic management.

For the applications reported, data fusion techniques appear promising. However, these encouraging results should not conceal the challenges that still remain. These include obtaining data with the necessary accuracy to create effective applications, dynamic and real-time issues associated with data quality as traffic flow changes, processing of data in real time, developing methods to combine sensor or hard data with human-generated or soft data, using measures of performance to assess fusion system effectiveness, and ensuring data fusion system security.[3,10,186,187]

References

1. N.-E. El Faouzi, H. Leung, and A. Kurian, "Data fusion in intelligent transportation systems: Progress and challenges–A survey," *Information Fusion* **12**(1), 4–10 (2011).
2. L. A. Klein, *Sensor and Data Fusion: A Tool for Information Assessment and Decision Making*, Second Edition, SPIE Press, Bellingham, Washington (2012) [doi: 10.1117/3.928035].
3. N.-E. El Faouzi and L. A. Klein, "Data Fusion in Intelligent Traffic and Transportation Engineering: Recent Advances and Challenges," in *Multisensor Data Fusion: From Algorithm and Architecture Design to Applications*, H. Fourati, Ed., Chapter 32, CRC Press, Boca Raton, Florida (2015).
4. N.-E. El Faouzi and L. A. Klein, "Data fusion for ITS: techniques and research needs," *International Symposium on Enhancing Highway Performance (ISEHP) 2016, Transportation Research Procedia* **15**, 495–512 (2016).
5. E. Waltz and J. Llinas, *Multisensor Data Fusion*, Artech House, Norwood, Massachusetts (1990).

6. F. E. White, Jr., "Joint Directors of Laboratories data fusion subpanel report: SIGINT session," *Tech. Proc. Joint Service Data Fusion Symposium* Vol. I, DFS-90, pp. 469–484 (1990).
7. J. Llinas, C. Bowman, G. Rogova, A. Steinberg, E. Waltz, and F. White, "Revisiting the JDL data fusion model II," *Proc. 7th International Conf. on Information Fusion*, Vol. II, Stockholm, Jun. 28–Jul. 1, 2004; P. Svensson and J. Schubert, International Society of Information Fusion, Mountain View, California, pp. 1218–1230, Jun. (2004).
8. A. N. Steinberg, C. L. Bowman, and F. E. White, Jr., "Revisions to the JDL data fusion model," *Proc. SPIE* **3719**, 430–441 (1999) [doi: 10.1117/12.341367]; also in A. N. Steinberg, C. L. Bowman, and F. E. White, Jr., "Revisions to the JDL model," *Joint NATO/IRIS Conference Proceedings*, Quebec, Oct. (1998).
9. A. N. Steinberg and C. L. Bowman, "Rethinking the JDL data fusion levels," *Proc. National Symposium on Sensor and Data Fusion (NSSDF)*, Johns Hopkins University Applied Physics Laboratory, Laurel, Maryland Jun. (2004).
10. B. Khaleghi, A. Khamis, F. O. Karray, and S. N. Razav, "Multisensor data fusion: A review of the state-of-the-art," *Information Fusion* **14**, 28–44 (2013).
11. F. Castanedo, "A review of data fusion techniques," *The Scientific World Journal* **2013**, 704504, 19 pages (2013).
12. O. Kessler, K. Askin, N. Beck, J. Lynch, F. White, D. Buede, D. Hall, and J. Llinas, *Functional Description of the Data Fusion Process*, Office of Naval Technology, Office of Naval Technology, Naval Air Development Center, Warminster, Pennsylvania (1991).
13. H. Boström, S. F. Andler, M. Brohede, R. Johansson, A. Karlsson, J. van Laere, L. Niklasson, M. Nilsson, A. Persson, and T. Ziemke, *On the Definition of Information Fusion as a Field of Research*, Univ. of Skövde Tech. Rpt. HS-IKI-TR-006, Skövde, Sweden (2007).
14. B. V. Dasarathy, "Information Fusion—What, where, why, when, and how?" *Information Fusion* **2**, 75–76 (2001).
15. N.-E. El Faouzi, "Multiform traffic data collection and data fusion in road traffic engineering," *Proc. Multisource Data Fusion in Traffic Engineering Workshop*, INRETS, Bron, France (2003).
16. *Connected Vehicle Pilot Deployment Program*, ITS Joint Program Office, U.S. Department of Transportation, Washington, D.C. (2015) http://www.its.dot.gov/pilots/index.htm. Accessed Nov. 18, 2015.
17. P. E. Ross, "Europe's smart highway will shepherd cars from Rotterdam to Vienna," *IEEE Spectrum*, 30 Dec. 2014. http://spectrum.ieee.org/transportation/advanced-cars/europes-smart-highway-will-shepherd-cars-from-rotterdam-to-vienna. Accessed Nov. 12, 2015.

18. J. L. Ygnace, *Travel Time/Speed Estimates on the French Rhone Corridor Network using Cellular Phones as Probes: STRIP Project*, European Community Research Program SERTI, INRETS-LESCOT, Lyon, France (2001).
19. Y. Youngbin and R. Cayford, *Investigation of Vehicles as Probes using Global Positioning System and Cellular Phone Tracking*, University of California PATH, Berkeley, California (2000).
20. L. A. Klein and M. R. Kelley, *Detection Technology for IVHS, Volume I: FHWA-RD-95-100*. Final Report, Federal Highway Administration, U.S. Department of Transportation, Washington, D.C. (1996).
21. L. A. Klein, *Sensor Technologies and Data Requirements for ITS*, Artech House, Boston (2001).
22. L. A. Klein, *ITS Sensors and Architectures for Traffic Management and Connected Vehicles*, CRC Press, Boca Raton, Florida (2017).
23. *Systems Engineering Guidebook for Intelligent Transportation Systems*, Version 3.0, U.S. Department of Transportation, Federal Highway Administration, California Division and California Department of Transportation, Nov. 2009. Available at http://www.fhwa.dot.gov/cadiv/segb/. Accessed Aug. 7, 2018.
24. R. Viswanathan and P. K. Varshney, "Distributed detection with multiple sensors: Part I—Fundamentals," *Proc. IEEE* **85**(1), 54–63 (1997).
25. M. E. Liggins II, C.-Y. Chong, I. Kadar, M. G. Alford, V. Vannicola, and S. Thomopoulos, "Distributed fusion architectures and algorithms for target tracking," *Proc. IEEE* **85**(1), 95–107 (1997).
26. B. V. Dasarathy, "Sensor fusion, potential exploitation: innovative architectures and illustrative applications," *Proc. IEEE* **85**(1), 24–38 (1997).
27. H. Durrant-Whyte and T. Henderson, "Multisensor data fusion," in *Handbook of Robotics, Part C—Sensing and Perception*, O. Khatib and B. Siciliano, Eds., Springer, New York, Chapter 25 (2008).
28. L. A. Klein, *Millimeter-Wave and Infrared Multisensor Design and Signal Processing*, Artech House, Norwood, Massachusetts (1997).
29. H. V. Kennedy, "Modeling second-generation thermal imaging systems," *Proc. SPIE* **1309**, 2–16 (1990) [doi:10.1117/12.21753].
30. E. Blasch, A. Steinberg, S. Das, J. Llinas, C.-Y. Chong, O. Kessler, E. Waltz, and F. White, "Revisiting the JDL model for information exploitation," *Proc. 16t Int'l Conf. on Information Fusion*, 9–12 July 2013, Instanbul (2013).
31. E. Blasch, "One decade of the Data Fusion Information Group (DFIG) model," *Proc. SPIE* **9499**, 94990L (2015) [doi: 10.1117/12.2176934].
32. "1989 data fusion survey," *Proc. 1990 Data Fusion Symposium*, Johns Hopkins University Applied Physics Laboratory, Laurel, Maryland, May (1990).

33. Data Fusion Development Strategy Panel, *Functional Description of the Data Fusion Process*, Office of Naval Technology, Warminster, Pennsylvania (1991).
34. D. L. Hall, *Mathematical Techniques in Multisensor Data Fusion*, Artech House, Norwood, Massachusetts (1992).
35. J. Llinas, "Data fusion overview," presented at the Center for Multisource Information Fusion, Univ. of Buffalo, Buffalo, New York, Oct. 2002.
36. F. E. White, "A model for data fusion," *Proc. 1st National Symposium on Sensor Fusion* (1988).
37. B. V. Dasarathy, "Information fusion, data mining, and knowledge discovery," *Information Fusion* **4**(1), 1 (2003).
38. E. P. Blasch and S. Plano, "JDL level 5 fusion model: user refinement issues and application in group tracking," *Proc. SPIE* **4729**, 270–279 (2002) [doi: 10.1117/12.477612].
39. E. P. Blasch and S. Plano, "Level 5: user refinement to aid the fusion process," *Proc. SPIE* **5099**, 288–297 (2003) [doi: 10.1117/12.486899].
40. M. Nilsson and T. Ziemke, "Rethinking Level 5: Distributed cognition and information fusion," *Proc. 9th International Conf. on Information Fusion*, Florence, 1–8, Jul., pp. 10–13 (2006).
41. A.-C. Boury-Brisset, "Ontological approach to military knowledge modeling and management," in *RTO-MP-IST-040*; presented at RTO 1st Symposium on Military Data and Information Fusion, Prague, Czech Republic, Oct. 20–22, 2003; also published by Defence R&D Canada Valcartier, Quebec, Mar. (2004).
42. E. Blasch, P. Valin, A.-L. Jousselme, D. Lambert, and E. Bosse, "Top ten trends in high-level information fusion," *Proc. 15th International Conference on Information Fusion* (FUSION 2012), pp. 2323–2330 (2012).
43. D. L. Hall and A. Steinberg, *Dirty Secrets in Multisensor Data Fusion*, The Pennsylvania State University Applied Research Laboratory, Jan. (2001) https://www.researchgate.net/publication/247936235_Dirty_Secrets_in_Multisensor_Data_Fusion. Accessed Sep. 18, 2018.
44. A. N. Steinberg, C. L. Bowman, and F. E. White, "Revisions to the JDL data fusion model," ERIM International, Arlington, Jan. 1999. http://www.dtic.mil/dtic/tr/fulltext/u2/a391479.pdf. Accessed Sep. 26, 2018.
45. D. L. Hall and R. J. Linn, "Algorithm selection for data fusion systems," *1987 Tri-Service Data Fusion Symposium Technical Proceedings*, Vol. I, DFS-87, Naval Air Development Center, Warminster, Pennsylvania (1987).
46. D. L. Hall and R. J. Linn, "A taxonomy of algorithms for multi-sensor data fusion," *Tech. Proc. Joint Service Data Fusion Symposium*, Vol. I, DFS-90, 594-610, Naval Air Development Center, Warminster, Pennsylvania (1990).

47. I. R. Goodman, R. P. S. Mahler, and H. T. Nguyen, *Mathematics of Data Fusion*, Kluwer Academic Publishers, Norwell, Massachusetts (1997).
48. Z. Ye, Y. Zhang, and D. R. Middleton, "An unscented Kalman filter method for speed estimation using single loop detector data," presented at 85th Annual Meeting of the Transportation Research Board, paper 06-2685, Washington, D.C. (2006).
49. P. Cheng, Z. Qiu, and B. Ran, "Traffic estimation based on particle filtering with stochastic state reconstruction using mobile network data," presented at 85th Annual Meeting of the Transportation Research Board, paper 06-2395, Washington, D.C. (2006); also appears in *Proc. IEEE Intelligent Transportation Systems Conf.*, Toronto, Ontario, Sep. 17–20 (2006) [doi: 10.1109/ITSC.2006.1707360].
50. J. Pearl, *Probabilistic Reasoning in Intelligent Systems: Networks of Plausible Inference*, Morgan Kaufmann Publishers, San Mateo, California (1988).
51. J. Dezert, "Foundations for a new theory of plausible and paradoxical reasoning," *Information and Security* **9**, 1–45 (2002).
52. C. K. Murphy, "Combining belief functions when evidence conflicts," *Decision Support Systems* **29**(1), 1–9 (2000).
53. P. Smets and R. Kennes, "The transferable belief model," *Artif. Intell.* **66**, 191–234 (1994).
54. B. R. Cobb and P. P. Shenoy, "A comparison of methods for transforming belief function models to probability models," in *Symbolic and Quantitative Approaches to Reasoning with Uncertainty*, T. D. Nielsen and N. L. Zhang Eds., *Lecture Notes in Artificial Intelligence*, Springer-Verlag, Berlin, pp. 255–266 (2003).
55. A. Jøsang, "The consensus operator for combining beliefs," *AI Journal* **14**(1–2), 157–170 (2002).
56. R. Haenni and N. Lehmann, "Probabilistic augmentation systems: a new perspective on Dempster–Shafer theory," *Int. J. of Intell. Sys.* **18**(1), 93–106 (2003).
57. D. Fixsen and R. P. S. Mahler, "The modified Dempster–Shafer approach to classification," *IEEE Trans. Sys., Man, and Cybern.–Part A: Systems and Humans (SMC)* **27**(1), 96–104 (1997).
58. A.-S. Capelle, C. Fernandez-Maloigne, and O. Colot, "Introduction of spatial information within the context of evidence theory," *IEEE International Conf. on Acoustics, Speech, and Sig. Proc. (ICASSP)*, pp. 785–788 (2003).
59. S. V. Hernandez, *Integration of Weigh-In-Motion and Inductive Signature Data for Truck Body Classification*, Ph.D. Thesis, University of California, Irvine Department of Civil and Environmental Engineering, (2014) https://escholarship.org/uc/item/73137048. Accessed Mar. 19, 2019.

60. S. Haykin, *Neural Networks: A Comprehensive Foundation*, Second Edition, Prentice Hall, Upper Saddle River, New Jersey (1998).
61. G. Wang and J. Kim, "A large-scale neural network model for real-time crash prediction in urban road networks," paper 18-06519, presented at 97th Annual Meeting of Transportation Research Board, Washington, D.C., Jan. 7–11 (2018).
62. C. Oh, J.-S. Oh, and S. G. Ritchie, "Real-time hazardous traffic condition warning system: framework and evaluation," *IEEE Transactions on Intelligent Transportation Systems* **6**(3), 265–272 (2005).
63. C. Oh and S. G. Ritchie, "Recognizing vehicle classification information from blade sensor signature," *Pattern Recognition Letters* **28**(9), 1041–1049 (2007).
64. S. V. Hernandez, A. Tok, and S. G. Ritchie, "Integration of weigh-in-motion (WIM) and inductive signature data for truck body classification," *Transportation Research Part C* **68**, 1–21 (2016).
65. I. U. Ahmed, R. Banik, M. M. Hasnat, M. Hadiuzzaman, T. Z. Qiu, and F. Rahman, "Probabilistic neural network and adaptive neuro fuzzy inference system based paratransit service quality predicting and attribute ranking," paper 17-00259, presented at 96th Annual Meeting of Transportation Research Board, Washington, D.C., Jan. 8–12 (2017).
66. E. Parzen, "On estimation of a probability density function and mode," *Ann. Math. Stat.* **33**(3), 1065–1076 (1962).
67. D. F. Specht, "Probabilistic neural networks and the polynomial adaline as complementary techniques for classification," *IEEE Transactions on Neural Networks* **1**(1), 111–121 (1990).
68. B. E. Hansen, *Lecture Notes on Nonparametrics*, University of Wisconsin, Spring 2009. https://www.ssc.wisc.edu/~bhansen/718/NonParametrics1.pdf. Accessed Aug. 21, 2018.
69. E. McDermott and S. Katagiri, "Minimum classification error via a Parzen window based estimate of the theoretical Bayes classification risk," *Proc. 12th IEEE Workshop on Neural Networks for Signal Processing*, Martigny, Switzerland, pp. 415–424 (2002).
70. K. Kim and K. Jang, "Spatiotemporal clustering of freeway crashes," paper 12-4238, presented at 91st Annual Meeting of Transportation Research Board, Washington, D.C., Jan. 22–26 (2012).
71. M. Allahviranloo, R. Regue, and W. Recker, "Pattern clustering and activity inference," paper 14-1274, presented at 93rd Annual Meeting of Transportation Research Board, Washington, D.C., Jan. 12–16 (2014).
72. M. Rinaldi and C. M. J. Tampère, "Dynamic traffic control clustering," paper 15-2870, presented at 94th Annual Meeting of Transportation Research Board, Washington, D.C., Jan. 11–15 (2015).

73. E. Ohn Bar, "Learning to detect vehicles by clustering appearance patterns," *IEEE Transactions on Intelligent Transportation Systems* **16**(5), 2511–2521 (2015).
74. J. Zhao, H. Xu, D. Wu, and H. Liu, "An artificial neural network to identify pedestrians and vehicles from roadside 360-degree LiDAR data," paper 18-03129, presented at 97th Annual Meeting of Transportation Research Board, Washington, D.C., Jan. 7–11 (2018).
75. R. Schalkoff, *Pattern Recognition: Statistical, Structural, and Neural Approaches*, John Wiley & Sons, New York (1992).
76. J. Roy, "Combining elements of information fusion and knowledge-based systems to support situation analysis," *Proc. SPIE* **6242**, 624202 (2006) [doi: 10.1117/12.663896].
77. L. A. Zadeh, *Fuzzy Sets and Systems*, North-Holland Press, Amsterdam (1978).
78. L. A. Zadeh, "Fuzzy logic," *IEEE Computer Magazine* **20**(4), 83–93 (1988).
79. B. Kosko, *Neural Networks and Fuzzy Systems: A Dynamical Systems Approach to Machine Intelligence*, Prentice-Hall, Englewood Cliffs, New Jersey (1992).
80. Y. Bar-Shalom and T. E. Fortmann, *Tracking and Data Association*, Academic Press, Orlando, Florida (1988).
81. M. P. Dana, "Registration: A prerequisite for multiple sensor tracking," in *Multitarget-Multisensor Tracking: Advanced Applications*, Y. Bar-Shalom, Ed., Artech House, Norwood, Massachusetts, Chapter 5 (1990).
82. S. S. Blackman, R. J. Dempster, and T. S. Nichols, "Application of multiple hypothesis tracking to multi-radar air defense systems," *Multi-Sensor Multi-Target Data Fusion, Tracking and Identification Techniques for Guidance and Control*, AGARD-AG-337, NATO (1996).
83. Y. Bar-Shalom, X. R. Li, and T. Kirubarajan, *Estimation with Applications to Tracking and Navigation: Theory, Algorithms, and Software*, John Wiley & Sons, New York (2001).
84. X. R. Li and V. P. Jilkov, "Survey of maneuvering target tracking: dynamic models," *Proc. SPIE* **4048**, 212–235 (2000) [doi: 10.1117/12.391979].
85. X. R. Li and V. P. Jilkov, "Survey of maneuvering target tracking: II. Ballistic target models," *Proc. SPIE* **4473**, 559–581 (2001) [doi: 10.1117/12.492753].
86. X. R. Li and V. P. Jilkov, "Survey of maneuvering target tracking: III. Measurement models," *Proc. SPIE* **4473**, 423–446 (2001) [doi: 10.1117/12.492752].
87. X. R. Li and V. P. Jilkov, "Survey of maneuvering target tracking: Decision-based methods," *Proc. SPIE* **4728**, 511–534 (2002) [doi: 10.1117/12.478535].

88. X. R. Li and V. P. Jilkov, "Survey of maneuvering target tracking—Part I: Dynamic models," *IEEE Trans. on Aero. and Elect. Sys.* **39**(4), 1333–1364 (2003).
89. X. R. Li and V. P. Jilkov, "A survey of maneuvering target tracking: approximation techniques for nonlinear filtering," *Proc. SPIE* **5428**, 537–550 (2004) [doi: 10.1117/12.553357].
90. X. R. Li and V. P. Jilkov, "Survey of maneuvering target tracking. Part V. Multiple-model methods," *IEEE Trans. on Aero. and Elect. Sys.* **41**(4), 1255–1321 (2005).
91. D. A. Lambert, "A blueprint for higher-level fusion systems," *Information Fusion* **10**(1), 6–24 (2009).
92. M. R. Endsley, "Design and evaluation for situation awareness enhancement," *Proc. Human Factors Society 32nd Annual Meeting*, Human Factors and Ergonomics Society, Santa Monica, pp. 97–101 (1988).
93. E. Bosse, A.-L. Jousselme, and P. Maupin, "Situation analysis for decision support: A formal approach," *Proc. 10th Int. Conf. on Info. Fusion*, Quebec, Jul. 9–12 (2007).
94. J. Roy, "From data fusion to situation analysis," *Proc. Fusion'2001 International Conference*, Montréal (2001).
95. E. Blasch, I. Kadar, J. Salerno, M. M. Kokar, S. Das, G. M. Powell, D. D. Corkill, and E. H. Ruspini, "Issues and challenges of knowledge representation and reasoning methods in situation assessment (Level 2 fusion)," *Journal of Advances in Information Fusion* **1**(2), 122–139 (2006).
96. G. Tadda, J. J. Salerno, D. Boulware, M. Hinman, and S. Gordon, "Realizing situation awareness in a cyber environment," *Proc. SPIE* **6242** (2006) [doi: 10.1117/12.665763].
97. K. Golestan, R. Soua, F. Karray, and M. S. Kamel, "Situation awareness within the context of connected cars: A comprehensive review and recent trends," *Information Fusion* **29**, 68–83 (2016).
98. K. Golestan, R. Soua, F. Karray, and M. S. Kamel, "A model for situation and threat/impact assessment in vehicular ad-hoc networks," DIVANet '14, *Proc. Fourth ACM International Symposium on Development and Analysis of Intelligent Vehicular Networks and Applications*, Montreal, Sep. 21–26, pp. 87–94 (2014).
99. K. Golestan, B. Khaleghi, F. Karray, and M. S. Kamel, "Attention assist: A high-level information fusion framework for situation and threat assessment in vehicular ad hoc networks," *IEEE Trans. on Intelligent Transportation Systems* **17**(5), 1271–1285 (2016).
100. A. N. Steinberg, "Data fusion system engineering," *Proc. 3rd International Conference on Information Fusion, FUSION 2000*, paper Mo05-3–MoD5-10, Jul. 10–13 (2000).
101. C. L. Bowman and A. N. Steinberg, "A Systems Engineering Approach for Implementing Data Fusion Systems," in *Handbook of Multisensor*

Data Fusion, D. L. Hall and J. Llinas, Eds., CRC Press, Boca Raton, Florida, Chapter 16 (2001).
102. A. Steinberg, "Problem-solving approach to data fusion," *Proc. 4th International Conference on Information Fusion*, International Society of Information Fusion, Montréal, Aug. 7–10 (2001).
103. M. Nilsson and T. Ziemke, *Investigating Human-Computer Interaction Issues in Information-Fusion-Based Decision Support*, Tech. Rpt. HS-IKI-TR-08-002, School of Humanities and Informatics, University of Skövde, Sweden, Jul. (2008).
104. R. Parasuraman, T. B. Sheridan, and C. D. Wickens, "A model for types and levels of human interaction with automation," *IEEE Trans. on Systems, Man and Cybernetics—Part A: Systems and Humans*, **30**(3), 286–297 (2000).
105. S. Snell, "The dissemination and fusion of geographical data to provide distributed decision making in a network-centric environment," *Proc. 9th International Command and Control Research and Technical Symp.*, Washington, D.C. (2004).
106. M. Nilsson and T. Ziemke, "Information fusion: A decision support perspective," *Proc. 10th International Conference on Information Fusion*, Quebec, July 9–12 (2007).
107. *ITS ePrimer*, Office of the Assistant Secretary for Research and Technology, U.S. Department of Transportation, Washington, D.C. (2014) http://www.pcb.its.dot.gov/ePrimer.aspx.
108. C. Taylor, D. Meldrum, and L. Jacobson, "Fuzzy ramp metering: Design overview and simulation results," *Transportation Research Record* **1634**, Transportation Research Board, Washington, D.C., pp. 10–18 (1998).
109. J. Niittymaki and S. Kikuchi, "Application of fuzzy logic to the control of a pedestrian crossing signal," *Transportation Research Record* **1651**, Transportation Research Board, Washington, D.C., pp. 30–38 (1998).
110. F. Koppelman, V. Sethi, and J. Ivan, *Calibration of Data Fusion Algorithm Parameters with Simulated Data, ADVANCE Project Report TRF-ID-152*, The Transportation Research Center, Northwestern University, Evanston, Illinois (1994).
111. J. N. Ivan, J. L. Schofer, F. S. Koppelman, and L. L. E. Massone, "Real-time data fusion for arterial street incident detection using neural networks," *Transportation Research Record* **1497**, Transportation Research Board, Washington, D.C., pp. 27–35 (1995).
112. J. N. Ivan, "Neural network representations for arterial street incident detection data fusion," *Transportation Research C: Emerging Technologies* **5**(34), 247–254 (1997).
113. S. C. Byun, D. B. Choi, B. H. Ahn, and H. Ko, "Traffic incident detection using evidential reasoning-based data fusion," *Proc. 6th World Congress on ITS*, ITS America, Toronto (1999).

114. N. E. Thomas, "Multi-sensor multivariate and multiclass incident detection system for arterial streets." *Proc. 16th International Symposium on Transportation and Traffic Theory (ISTTT)*, Lyon, France, pp. 315–340 (1996).
115. L. A. Klein, "Dempster–Shafer data fusion at the traffic management center," presented at 79th Annual Meeting of the Transportation Research Board, paper 00-1211, Washington, D.C. (2000).
116. L. A. Klein, P. Yi, and H. Teng, "Decision support system for advanced traffic management through data fusion," *Transportation Research Record* **1804**, Transportation Research Board, Washington, D.C., pp. 173–178 (2002).
117. S. Cohen, "Fusion of incident detection algorithms," *Proc. Multisource Data Fusion in Traffic Engineering Workshop, Les collections de l'INRETS*, Bron, France, pp. 109–117 (2003).
118. J. Mueck, "Estimation methods for the state of traffic at traffic signals using detectors near the stop-line," *Traffic Engineering and Control* **43**(11), (2002).
119. B. Friedrich and R. Minciardi, "Data fusion techniques for adaptive traffic signal control," *IFAC Symposium on Transportation Systems*, International Federation of Automatic Control, Tokyo, pp. 53–55 (2003).
120. Y. Wang and M. Papageorgiou, "Real-time freeway traffic state estimation based on extended Kalman filter: A general approach," *Transportation Research B* **39**, 141–147 (2005).
121. L. Li, S. He, F. Yang, J. Zhang, and B. Ran, "Space-mean-speed prediction based on the fusion of multi-source data," presented at 79th Annual Meeting of the Transportation Research Board, paper 17-03771, Washington, D.C. (2017).
122. Q.-J. Kong, Z. Li, Y. Chen, and Y. Liu, "An approach to urban traffic state estimation by fusing multisource information," *IEEE Trans. on Intelligent Transportation Systems* **10**(3), 499–511 (2009).
123. T. Denoeux, "A *k*-nearest neighbor classification rule based on Dempster–Shafer theory," *IEEE Trans. on Systems, Man, and Cybernetics* **25**(5), 804–813 (1995).
124. N. A. Carlson, "Federated square root filter for decentralized parallel process," *IEEE Trans. Aerosp. Electron. Syst.* **26**(3), 517–525 (1990).
125. N. A. Carlson and M. P. Berarducci, "Federated Kalman filter simulation results," *Navigation* **41**(3), 297–322 (1994).
126. J. Gao, K. Ozbay, and A. Kurkcu, "Multi-source data fusion for urban traffic state estimation: a case study of New York City," presented at *98th* Annual Meeting of the Transportation Research Board, paper 19-04253, Washington, D.C., Jan. 13–17 (2019).
127. *Data Fusion for Delivering Advanced Traveler Information Services*, ITS Joint Program Office, Federal Highway Administration, Washington, D.C., pp. 247–254 (2003).

128. A. Tarko and N. Rouphail, "Travel time fusion in ADVANCE," presented at the Pacific Rim TransTech Conference: A Ride into the Future, Seattle (1993).
129. P. Thakuriah and A. Sen, *Data Fusion for Travel Time Prediction: A Statement of Requirements, ADVANCE Working Paper Series* **22**, *Task TRF-TT-07*, Urban Transportation Center, University of Illinois, Chicago (1993).
130. S. Berka, X. Tian, and A. Tarko, *Data Fusion Algorithm for ADVANCE, ADVANCE Working Paper Series* **48**, Urban Transportation Center, University of Illinois, Chicago (1995).
131. D. J. Dailey, P. Harn, and P.-J. Linet, *ITS Data Fusion, Research Project T9903, Task 9: ATIS/ATMS Regional IVHS Demonstration, Report WA-RD 410.1*, Washington State Department of Transportation, Seattle (1996).
132. A. Abe, "The correction of the forecasting travel time by using AVI data," *Proc. IVHS* (1997).
133. R. D. Kühne, "Data fusion for DRG systems," *IFAC Symposium on Transportation Systems* **3**, International Federation of Automatic Control, Chania, Greece (1997).
134. K. Choi, "Data fusion for generating the link travel time with insufficient data sources," *Proc. IVHS* (1997).
135. K. Y. Choi and Y. Chung, "Travel time estimation algorithm using GPS probe and loop detector data fusion," presented at 80th Annual Meeting of the Transportation Research Board, paper 01-0374, Washington, D.C. (2001).
136. N.-E. El Faouzi and J.-B. Lesort, "Travel time estimation on urban networks from traffic data and on-board trip characteristics," *Proc. Second World Congress on ITS*, Morgan Kaufmann Publishers, Yokohama, pp. 88–93 (1995).
137. N.-E. El Faouzi, "Heterogeneous data source fusion for impedance indicators," *IFAC Symposium on Transportation Systems*, Chania, Greece, pp. 1375–1380 (1997).
138. N.-E. El Faouzi, "Bayesian and evidential approaches for traffic data fusion: methodological issues and case study," presented at 85th Annual Meeting of the Transportation Research Board, paper 06-1510, Washington, D.C. (2006).
139. N.-E. El Faouzi, "Travel time estimation via evidential data fusion," *Recherche Transports Sécurité* **68**, 15–30 (2000).
140. N.-E. El Faouzi, L. A. Klein, and O. De Mouzon, "Improving travel time estimates from inductive loop and toll collection data with Dempster–Shafer data fusion," *Transportation Research Record* **2129**, Transportation Research Board, pp. 73–80 (2009).

141. X. Zhang, M. Hamedi, and A. Haghani, "A real-time data fusion framework for corridor travel time estimation with multiple data sources," presented at 79th Annual Meeting of the Transportation Research Board, paper 17-04974, Washington, D.C. (2017).
142. F. Dion and H. Rakha, "Estimating dynamic roadway travel times using automatic vehicle identification data for low sampling rates," *Transportation Research Part B: Methodological* **40**(9), 745–766 (2006).
143. H. Durrant-Whyte and T. Bailey, "Simultaneous localization and mapping (SLAM): Part I–The essential algorithms," *Robotics and Automation Magazine* **13**, 99–110 (2006).
144. R. R. Murphy, "Sensor and information fusion for improved vision-based vehicle guidance," *IEEE Intelligent Systems* **13**(6), 49–56 (1998).
145. S.-C. Pei and G. Liou, "Vehicle-type motion estimation by the fusion of image point and line features," *Pattern Recognition* **31**, 333–344 (1998).
146. J. Langheim and A. J. Buchanan, "CARSENSE—Sensor fusion for DAS," *Proc. ITS in Europe* (2002).
147. C. Stiller, J. Hipp, C. Rossig, and A. Ewald, "Multisensor obstacle detection and tracking," *Proc. IEEE International Conference on Intelligent Vehicles*, pp. 451–456 (1998).
148. M. Tsogas, X. Dai, G. Thomaidis, P. Lytrivis, and A. Amditis, "Detection of maneuvers using evidence theory," *Proc. IEEE Intelligent Vehicles Symp.*, Eindhoven University of Technology, Eindhoven, Jun. 4–6 (2008).
149. R. R. Yager, "On the Dempster–Shafer framework and new combination rules," *Information Science* **41**(2), 93–137 (1987).
150. S.-Y. Sohn and H. W. Shin, "Data mining for road traffic accident classification," *Ergonomics* **44**, 107–117 (2001).
151. S.-Y. Sohn and S. H. Lee, "Data fusion, ensemble clustering to improve the classification accuracy for the severity of road traffic accidents in Korea," *Safety Science* **41**, 1–14 (2003).
152. L. Breiman, "Bagging predictors," *Machine Learning* **26**, 123–140 (1996).
153. L. Breiman, "Arcing classifiers," *Annals of Statistics* **26**, 801–849 (1998).
154. L. Breiman, "Random forests," *Machine Learning* **45**, 5–32 (2001).
155. D. H. Wolpert, "Stacked generalization," *Neural Networks* **5**, 241–259 (1992).
156. M. Cremer and H. Keller, "Dynamic identification of OD flows from counts at complex intersections," *Proc. 8th International Symposium on Transportation and Traffic Theory (ISTTT)*, University of Toronto Press, Toronto (1981).
157. M. Cremer, "Determining time dependent trip distribution in a complex intersection for traffic responsive control," *Proc. Control in*

Transportation Systems, IFAC-IFIP-IFORS, International Federation of Automatic Control, Baden-Baden, Germany (1983).

158. I. Okutani, "The Kalman filtering approaches in some transportation and road traffic problems," *Proc. 10th International Symposium on Transportation and Traffic Theory (ISTTT)*, Waltham, Massachusetts, pp. 397–416 (1987).
159. K. Ashok and M. Ben-Akiva, "Dynamic O–D matrix estimation and prediction for real-time management systems," *Proc. 12th International Symposium on Transportation and Traffic Theory (ISTTT)*, Berkeley, pp. 465–484 (1993).
160. M. Ben-Akiva and T. Morikawa, "Methods to combine different data sources and estimates OD matrices," *Proc. 10th International Symposium on Transportation and Traffic Theory (ISTTT)*, Cambridge, Massachusetts, pp. 459–481 (1987).
161. J. T. Lundgren, A. Peterson, and S. Tengroth, "Methods for pre-adjusting time dependent origin–destination matrices," *Proc. 10th World Congress and Exhibition on Intelligent Transportation Systems and Services*, Madrid (2003).
162. K. Hyun, A. Tok, and S. G. Ritchie, "Long distance truck tracking from advanced point detectors using a selective weighted Bayesian model," *Transportation Research Part C: Emerging Technologies* **82**, 24–42 (2017).
163. J. F. Harrison and F. Stevens, "A Bayesian approach to short-term forecasting," *Operational Research Quarterly* **22**, 341–362 (1971).
164. I. Okutani and J. Stephanedes, "Dynamic prediction of traffic volume through Kalman filtering theory," *Transportation Research B* **18**, 1–11 (1984).
165. P. C. Vythoulkas, "Alternative approaches to short term traffic forecasting for use in driver information systems," *Proc. 12th International Symposium on Transportation and Traffic Theory (ISTTT)*, Berkeley, pp. 485–506 (1993).
166. N.-E. El Faouzi, "Nonparametric traffic flow prediction using kernel estimator," *Proc. 13th International Symposium on Transportation and Traffic Theory (ISTTT)*, Lyon, France, pp. 41–54 (1996).
167. J. M. Bates and C. W. J. Granger, "The combination of forecasts," *Operational Research Quarterly* **20**(4), 51–468 (1969).
168. W. J. Granger, "Combining forecasts: Twenty years later," *Journal of Forecasting* **8**, 167–173 (1989).
169. N.-E. El Faouzi, "Combining predictive schemes in short-term traffic forecasting," *Proc. 17th International Symposium on Transportation and Traffic Theory (ISTTT)*, Jerusalem, Israel, pp. 471–487 (1999).
170. M. Cremer and S. Schrieber, "Monitoring traffic load profiles with heterogeneous data sources configurations," *Proc. 13th International*

Symposium on Transportation and Traffic Theory (ISTTT), Lyon, France, pp. 21–40 (1996).
171. J. Sau, N.-E. El Faouzi, A. Ben Aïssa, and O. De Mouzon, "Particle filter-based realtime estimation and prediction of traffic conditions," *Proc. Applied Stochastic Models and Data Analysis (ASMDA) International Society*, ASMDA International, Chania, Greece (2007).
172. D. B. Choi, "On multisensor data fusion using attributes association for intelligent traffic congestion information inference," *Proc. IVHS* (1989).
173. P. Sekula, N. Markovic, Z. Vander Laana, and K. F. Sadabadi, "Estimating historical hourly traffic volumes via machine learning and vehicle probe data: A Maryland case study," *Transportation Research Part C: Emerging Technologies* **97**, 147–158 (2018).
174. P. Domingos and M. Pazzani, "On the optimality of the simple Bayesian classifier under zero-one loss," *Machine Learning* **29**(2–3), 103–130 (1997).
175. M. S. Grewal, L. R. Weill, and A. P. Andrews, *Inertial Navigation and Integration*, John Wiley & Sons, Hoboken, New Jersey (2004).
176. M. Wei and K. P. Schwarz, "Testing a decentralized filter for GPS/INS integration," *Proc. IEEE Position Location and Navigation Symposium*, Las Vegas, pp. 429–435 (1990).
177. W. Li and H. Leung, "Constrained unscented Kalman filter based fusion of GPS/INS/digital map for vehicle localization," *IEEE 2003 International Conference on Intelligent Transportation Systems*, Shanghai, China, pp. 362–1367 (2003).
178. N. El-Sheimy, K. Chiang, and A. Noureldin, "The utilization of artificial neural networks for multi-sensor system integration in navigation and positioning instruments," *IEEE Transactions on Instrumentation and Measurement* **55**, 1606–1615 (2006).
179. R. Sharaf, A. Noureldin, A. Osman, and N. El-Sheimy, "Implementation of online INS/GPS integration utilizing radial basis function neural network," *IEEE Aerospace and Electronic Systems Magazine* **20**, 8–14 (2005).
180. A. Hiliuta, R. Landry, and F. Gagnon, "Fuzzy corrections in a GPS/INS hybrid navigation system," *IEEE Transactions on Aerospace and Electronic Systems* **2**, 591–600, 2004.
181. R. Toledo-Moreo and M. A. Zamora-Izquierdo, "IMM-based lane-change prediction in highways with low-cost GPS/INS," *IEEE Trans. on Intelligent Transportation Systems* **10**(1), 180–185 (2009).
182. H.-S. Tan and J. Huang, "DGPS-based vehicle-to-vehicle cooperative collision warning: Engineering feasibility viewpoints," *IEEE Trans. on Intelligent Transportation Systems* **7**(4), 415–428 (2006).
183. U. S. Air Force, "New Civil Signals," http://www.gps.gov/systems/gps/modernization/civilsignals/. Accessed Aug. 2, 2018.

184. J. Y. Kim, National Coordination Office for Space-Based PNT, personal communication, May 10, 2019.
185. I. Patey, "Use of big data for managing England's National Network," *International Symposium on Enhancing Highway Performance (ISEHP) 2016*, Berlin (2016).
186. R. Falcon, R. Abielmona, S. Billins, and A. Plachkov, "Risk management with hard-soft data fusion in maritime domain awareness," *Seventh IEEE Symposium on Computational Intelligence for Security and Defense Applications (CISDA)*, Dec. 14–17 (2014).
187. V. Dragos and S. Gatepaille, "On-the-fly integration of soft and sensor data for enhanced situation assessment," 21st International Conference on Knowledge Based and Intelligent Information and Engineering Systems—KES 2017, *Procedia Computer Science* **112**, 1263–1272 (2017).

Chapter 3
Bayesian Inference for Traffic Management

Bayesian inference is widely applied for detection, classification, and identification of objects and events that occur in conjunction with traffic management strategies. The applications include travel-time estimation, automatic incident detection, and decision support. This chapter describes the underlying principles of the technique, the assumptions typically encountered in its applications, and four examples that illustrate its exploitation for traffic management. The examples presented are identification of vehicle type using emission spectra data collected by a sensor; incident detection using an influence diagram and joint sensor report with data from roadway, cell phone, and truck radio reports; truck classification using a multiple-classifier system; and maneuver-based trajectory prediction and criticality assessment through causal Bayesian networks.

3.1 Bayesian Inference

Bayesian inference is a probability-based reasoning discipline grounded in Bayes' theorem. When used to support data fusion, Bayesian inference belongs to the class of parametric algorithms that use *a priori* (prior) knowledge about events or objects in an observation space to make inferences about the identity of events or objects in that space. Bayesian inference provides a method for calculating the conditional *a posteriori* (posterior) probability of a hypothesis being true given supporting evidence. Thus, Bayes' theorem offers a method for updating beliefs in response to information or evidence that would cause the belief to change.

Another interpretation of Bayesian inference is presented in this scenario. Suppose that it is necessary to determine the likelihoods of different values H_i of an unknown state. There may be prior beliefs about which H_i might be expected. These are encoded in the form of relative likelihoods in the prior probability $P(H_i)$. An observation E is made to obtain additional information about the particular H_i. The observations are modeled as a conditional probability $P(E|H_i)$ that describes, for each H_i, the probability of E given that a particular H_i is true. The new likelihoods associated with the H_i are computed from the product of the original prior information and the information gained by observation (the evidence). This is encoded in the posterior probability $P(H_i|E)$ that describes the likelihoods associated with H_i given the evidence E. In a fusion process, the factor $P(E)$ that appears in the denominator of Bayes' theorem is used to normalize the posterior and is not generally evaluated using its formal definition as the sum of the products of the likelihood functions and the prior probabilities over all hypotheses.[1–3]

3.2 Derivation of Bayes' Theorem

Bayes' theorem may be derived by evaluating the probability of occurrence of an arbitrary event E, assuming that another event H has occurred. The probability is given by

$$P(E|H) = \frac{P(EH)}{P(H)}, \tag{3.1}$$

where H is an event with positive probability. The quantity $P(E|H)$ is the probability of E conditioned on the occurrence of H. The conditional probability is not defined when H has zero probability. The factor $P(EH)$ represents the probability of the intersection of events E and H.

To illustrate the meaning of Eq. (3.1), consider a population of N people that includes N_E left-handed people and N_H females, as shown in the Venn diagram of Fig. 3.1.[4] Let E and H represent the events that a person chosen at random is left-handed or female, respectively. Then,

$$P(E) = \frac{N_E}{N}, \text{ and} \tag{3.2}$$

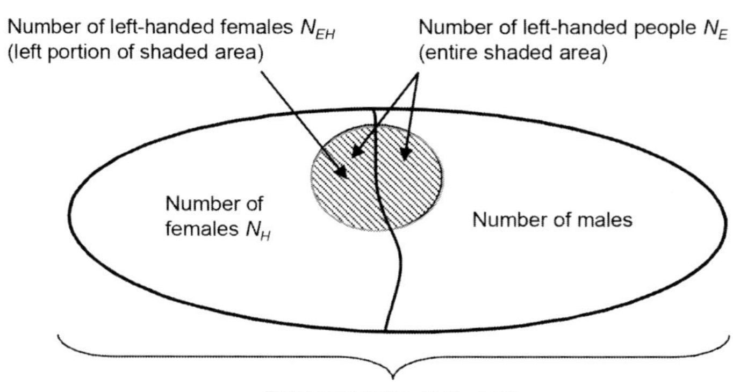

Figure 3.1 Venn diagram illustrating intersection of events E (person chosen at random is left-handed) and H (person chosen at random is female).

$$P(H) = \frac{N_H}{N}. \tag{3.3}$$

The probability that a female chosen at random is left-handed is N_{EH}/N_H, where N_{EH} is the number of left-handed females. In this example, $P(E|H)$ denotes the probability of selecting a left-handed person at random, assuming that the person is female. In terms of population parameters, $P(E|H)$ is

$$P(E|H) = \frac{N_{EH}}{N_H} = \frac{P(EH)}{P(H)}. \tag{3.4}$$

Returning to the derivation of Bayes' theorem, Eq. (3.1) may be rewritten as

$$P(EH) = P(E|H)P(H), \tag{3.5}$$

which is referred to as the theorem on compound probabilities.

When H consists of a set of mutually exclusive and exhaustive hypotheses $H_1, \ldots, H_n$, conditional probabilities, which may be easier to evaluate than unconditional probabilities, can be substituted for $P(EH)$ as follows. The mutually exhaustive property implies that one hypothesis necessarily is true; i.e., the union of $H_1, \ldots, H_n$ is the entire sample space. Under these conditions, any event E can occur only in conjunction with some H_j such that

$$E = EH_1 \cup EH_2 \cup \ldots \cup EH_n. \tag{3.6}$$

Since the EH_j are mutually exclusive, their probabilities add as

$$P(E) = \sum_{i=1}^{n} P(EH_i). \tag{3.7}$$

Upon substituting H_j for H and summing over i, Eq. (3.5) becomes

$$P(E) = \sum_{i} [P(E|H_i)P(H_i)] \tag{3.8}$$

when the identity in Eq. (3.7) is applied. Equation (3.8) states that the belief in any event E is a weighted sum over all of the distinct ways that E can be realized.

In Bayesian inference, we are interested in the probability that hypothesis H_i is true given the existence of evidence E. This statement is expressed as

$$P(H_i|E) = \frac{P(EH_i)}{P(E)}. \tag{3.9}$$

If Eqs. (3.5) and (3.8) are introduced into Eq. (3.9), Eq. (3.9) takes the following form of Bayes' theorem:

$$P(H_i|E) = \frac{P(E|H_i)P(H_i)}{P(E)} = \frac{P(E|H_i)P(H_i)}{\sum_i [P(E|H_i)P(H_i)]}, \tag{3.10}$$

where:
$P(H_i|E)$ = *a posteriori* probability that hypothesis H_i is true given evidence E,
$P(E|H_i)$ = probability of observing evidence E given that H_i is true (referred to as the likelihood function),
$P(H_i)$ = *a priori* probability that hypothesis H_i is true,

$$\sum_i P(H_i) = 1, \text{ and} \tag{3.11}$$

$\sum_i P(E|H_i)P(H_i)$ = preposterior or the sum of the products of the likelihood functions and the *a priori* probabilities over all hypotheses i.

To summarize, Bayes' theorem simply states that the posterior probability equals the product of the likelihood function and the prior probabilities divided by the evidence.

The likelihood functions represent the extent to which the posterior probability is subject to change. These functions are evaluated through offline experiments or by analyzing the available information for the problem at hand. A general method of estimating the parameter(s) that maximize the likelihood function given the data is to find the maximum-likelihood estimate. This procedure selects the parameter value that makes the data actually observed as likely as possible.[5–7] The preposterior is the sum of the products of the likelihood functions and the *a priori* probabilities. It can also be calculated from its role as a normalizing factor.[8] The approach used to compute its numerical value depends on the given information for a particular problem and the type of problem.

3.3 Likelihood Function and Prior Probability Models

The normal distribution is often used to model the likelihood function $P(E|H_i)$ and the prior probability $P(H_i)$.[1,2] Assuming that E and H are random variables, the likelihood function becomes

$$P(E|H) = \frac{1}{\sqrt{2\pi\sigma_E^2}} \exp\left[-\frac{1}{2}\frac{(E-H)^2}{\sigma_E^2}\right]. \tag{3.12}$$

Two possibilities exist for interpreting Eq. (3.12) since it is a function of two variables E and H. If the true value of the hypothesis $H = H_i$ is known, then

Eq. (3.12) is a function of the evidence or observation E only, which describes the probability of observing a particular hypothesis (normally distributed around the true hypothesis H_i with variance σ_E^2). Conversely, if we make a specific observation $E = E_k$ at time k, then the distribution is a function of H alone, which describes the probability of observing the true hypothesis value (normally distributed around the observation E with variance σ_E^2). In this case, the distribution is the likelihood function.

If a prior belief concerning the true state H exists such that the prior probability can be described with a normal distribution, then the prior probability becomes

$$P(H) = \frac{1}{\sqrt{2\pi\sigma_H^2}} \exp\left[-\frac{1}{2}\frac{(H - H_i)^2}{\sigma_H^2}\right]. \tag{3.13}$$

Equation (3.13) is a function of a single variable H since H_i is considered a constant. The prior information and observation evidence from the likelihood function can now be combined using Bayes' theorem to give

$$\begin{aligned} P(H|E) = {} & \frac{1}{P(E)} \frac{1}{\sqrt{2\pi\sigma_E^2}} \exp\left[-\frac{1}{2}\frac{(E - H)^2}{\sigma_E^2}\right] \\ & \times \frac{1}{\sqrt{2\pi\sigma_H^2}} \exp\left[-\frac{1}{2}\frac{(H - H_i)^2}{\sigma_H^2}\right] \end{aligned} \tag{3.14}$$

$$= \frac{1}{\sqrt{2\pi\sigma^2}} \exp\left[-\frac{1}{2}\frac{(H - \bar{H})^2}{\sigma^2}\right], \tag{3.15}$$

where $P(E)$ is treated as a normalizing factor, and $\bar{H}$ and σ^2 are given, respectively, as

$$\bar{H} = \frac{\sigma_H^2}{\sigma_H^2 + \sigma_E^2} E + \frac{\sigma_E^2}{\sigma_H^2 + \sigma_E^2} H_i \tag{3.16}$$

and

$$\sigma^2 = \frac{\sigma_E^2 \sigma_H^2}{\sigma_E^2 + \sigma_H^2} = \left(\frac{1}{\sigma_E^2} + \frac{1}{\sigma_H^2}\right)^{-1}. \tag{3.17}$$

Thus, the posterior probability is also normally distributed with a mean equal to the weighted average of the means of the original likelihood and prior, and a variance equal to the parallel combination of the original variances.

3.4 Monty Hall Problem

The classical Monty Hall problem from the game show Let's Make a Deal® describes "gifts" hidden behind three doors. Only one of the doors hides a valuable gift, such as an automobile, while the other two hide less desirable gifts such as goats. In the first formulation of the problem, Monty knows what's behind each door. This is critical information, as shown later. Monty asks the contestant to select the door that he or she thinks is hiding the valuable gift. Suppose that the contestant chooses Door 1 initially. Monty then reveals the goat located behind Door 2 or Door 3, as depicted in Fig. 3.2. The contestant is then asked if he or she wants to switch his or her door selection. Is it to the advantage of the contestant to switch or not? The problem can be solved using any of three methods.

3.4.1 Case-by-case analysis solution

In the first approach, a case-by-case analysis, a truth table confirms that the odds of winning the automobile if the contestant does not switch are 1:3 as only one of the three doors hides the automobile. Table 3.1 shows where the two goats (Billy and Millie) may be hidden by any two of the three doors. Monty will always reveal a goat, never the more valuable automobile.

The odds that the contestant will win the automobile by switching doors are described in Table 3.2. The tally of the case-by-case analysis is four *Yes* and two *No*. Thus, the odds are 2:1 in favor of switching after Monty Hall reveals the goat.

Figure 3.2 Monty Hall problem showing a goat behind one of three doors. In this example, there is a second goat behind either Door 1 or Door 2.

Table 3.1 Monty Hall problem with car behind one door and goats Billy or Millie behind the other two.

Door 1	Door 2	Door 3
Car	Billy	Millie
Car	Millie	Billy
Billy	Car	Millie
Millie	Car	Billy
Billy	Millie	Car
Millie	Billy	Car

Table 3.2 Case-by-case analysis of Monty Hall problem.

Case	Explanation
1	Monty Hall reveals a goat behind either Door 2 or Door 3. It is *not* to the contestant's advantage to switch. Record *No*.
2	Similar to case 1. It is *not* to the contestant's advantage to switch. Record *No*.
3	Monty Hall reveals a goat behind Door 3. It *is* to the contestant's advantage to switch. Record *Yes*.
4	Similar to case 3. It *is* to the contestant's advantage to switch. Record *Yes*.
5	Monty Hall reveals a goat behind Door 2. It *is* to the contestant's advantage to switch. Record *Yes*.
6	Similar to case 5. It *is* to the contestant's advantage to switch. Record *Yes*.

3.4.2 Conditional probability solution

The second way of solving the Monty Hall problem is by applying conditional probabilities. Let the three doors be denoted by 1, 2, and 3, and let C_1 be the event that the car is behind Door 1, C_2 be the event that the car is behind Door 2, and C_3 be the event that the car is behind Door 3.

Let H_1 represent the event that Monty Hall opens Door 1, H_2 represent the event that Monty Hall opens Door 2, and H_3 represent the event that Monty Hall opens Door 3.

Assuming that the game show host knows what item is behind each door and, furthermore, that the contestant chooses Door 1 to begin the game, the probability that the contestant wins a car if he or she then switches his or her choice is given by

$$\begin{aligned} P(H_3 \cap C_2) + P(H_2 \cap C_3) &= P(C_2)P(H_3|C_2) + P(C_3)P(H_2|C_3) \\ &= (1/3)(1) + (1/3)(1) = 2/3. \end{aligned} \tag{3.18}$$

The values of $P(H_3|C_2)$ and $P(H_2|C_3)$ are unity since Monty will always reveal a goat, as he knows what item is behind each door.

If the game show host does *not* know what item is behind each door or he forgets, then the probability of the contestant winning the car does not change if he or she switches his or her choice of doors. Mathematically, the probabilities become

$$\begin{aligned} P(H_3 \cap C_2) + P(H_2 \cap C_3) &= P(C_2)P(H_3|C_2) + P(C_3)P(H_2|C_3) \\ &= (1/3)(1/2) + (1/3)(1/2) = 1/3. \end{aligned} \tag{3.19}$$

In this case, Monty makes a random choice between the two remaining doors, and $P(H_3|C_2)$ and $P(H_2|C_3)$ each equal ½.

A website demonstration of the Monty Hall problem for both situations is found at www.math.ucsd.edu/~crypto/Monty/monty.html, and further discussions of the problem are offered at www.cut-the-knot.org/hall.shtml.

3.4.3 Bayesian inference solution

In Bayesian terms, a probability $P(A|I)$ is a number in |0, 1| associated with a proposition A. The number expresses a degree of belief in the truth of A subject to whatever *background* information I happens to be known.

For this problem, the background is provided by the rules of the game. The propositions of interest are:

C_i: The car is behind Door i, for i equal to 1, 2, or 3.
H_{ij}: The host opens Door j after the player has picked Door i, for i and j equal to 1, 2, or 3.

For example, C_1 denotes the proposition *the car is behind Door* 1, and H_{12} denotes the proposition *the host opens Door* 2 *after the player has picked Door* 1.

The assumptions underlying the common interpretation of the Monty Hall puzzle are then formally stated as follows.

First, the car can be behind any door, and all doors are *a priori* equally likely to hide the car. In this context, *a priori* means *before the game is played*, or *before seeing the goat*. Hence, the prior probability of a proposition C_i is

$$P(C_i|I) = \tfrac{1}{3}. \tag{3.20}$$

Second, the host will always open a door that has no car behind it chosen from among the two not picked by the player. If two such doors are available, each one is equally likely to be opened. This rule determines the conditional probability of a proposition H_{ij} subject to where the car is, i.e., *conditioned* on a proposition C_k according to

$$P(H_{ij}|C_k, I) = \begin{cases} 0 \text{ if } i = j \text{ (the host cannot open the door picked by the player)} \\ 0 \text{ if } j = k \text{ (the host cannot open a door with a car behind it)} \\ 1/2 \text{ if } i = k \text{ (the two doors with no car are equally likely to be opened)} \\ 1 \text{ if } i \neq k \text{ and } j \neq k \text{ (there is only one door available to open).} \end{cases} \tag{3.21}$$

The problem can now be solved by scoring each strategy with its associated posterior probability of winning, i.e., with its probability subject to the host's opening of one of the doors. Without loss of generality, assume, by renumbering the doors if necessary, that the player picks Door 1 and that the host then opens Door 3, revealing a goat. In other words, the host *makes* proposition H_{13} true.

The posterior probability of *winning by not switching* doors, subject to the game rules and H_{13}, is then $P(C_1|H_{13}, I)$. Bayes' theorem expresses this probability as

$$P(C_1|H_{13}, I) = \frac{P(H_{13}|C_1, I)P(C_1|I)}{P(H_{13}|I)}. \tag{3.22}$$

Using the above assumptions, the numerator of the right-hand side becomes

$$P(H_{13}|C_1, I)P(C_1|I) = \frac{1}{2} \times \frac{1}{3} = \frac{1}{6}. \tag{3.23}$$

The normalizing factor in the denominator is evaluated by expanding it using the definitions of marginal probability and conditional probability. Thus,

$$\begin{aligned} P(H_{13}|I) &= \sum_i P(H_{13}, C_i|I)P(C_i|I) \\ &= P(H_{13}, C_1|I)P(C_1|I) + P(H_{13}, C_2|I)P(C_2|I) \\ &\quad + P(H_{13}, C_3|I)P(C_3|I) \\ &= \frac{1}{2} \times \frac{1}{3} + 1 \times \frac{1}{3} + 0 \times \frac{1}{3} = \frac{1}{2}. \end{aligned} \tag{3.24}$$

Dividing the numerator by the normalizing constant yields

$$P(C_1|H_{13}, I) = \frac{1}{6} \div \frac{1}{2} = \frac{1}{3},$$

which equals the prior probability of the car being behind the initially chosen door, meaning that the host's action has not contributed any novel information with regard to this eventuality. In fact, the following argument shows that the effect of the host's action consists entirely of redistributing the probabilities for the car being behind either of the *other* two doors.

The probability of *winning by switching* the selection to Door 2, $P(C_2|H_{13}, I)$, is evaluated by requiring that the posterior probabilities of all of the C_i propositions add to 1. That is,

$$1 = P(C_1|H_{13}, I) + P(C_2|H_{13}, I) + P(C_3|H_{13}, I). \tag{3.25}$$

There is no car behind Door 3 since the host opened it, so the last term must be zero. This is proven using Bayes' theorem and the previous results as

$$P(C_3|H_{13}, I) = \frac{P(H_{13}|C_3, I)P(C_3|I)}{P(H_{13}|I)} = \left(0 \times \frac{1}{3}\right) \div \frac{1}{2} = 0. \tag{3.26}$$

Hence,

$$P(C_2|H_{13}, I) = 1 - \frac{1}{3} - 0 = \frac{2}{3}. \tag{3.27}$$

Equation (3.27) shows that the winning strategy is to switch the selection to Door 2. It also makes clear that the host's showing of the goat behind Door 3 has the effect of *transferring* the ⅓ of winning probability, *a priori* associated with that door, to the remaining unselected and unopened one, thus making it the most likely winning choice.

3.5 Application of Bayes' Theorem to Cancer Screening

Suppose that a patient visits a physician who proceeds to administer a low-cost screening test for cancer. The test has an accuracy of 95% (i.e., the test will indicate positive 95% of the time if the patient has the disease) with a 4% false-alarm probability. Furthermore, suppose that cancer occurs in 5 out of every 1,000 people in the general population. If the patient is informed that the test is positive for cancer, what is the probability that the patient actually has cancer?

The truth diagram in Fig. 3.3 summarizes the statistics for this example in terms of H_0 (patient does not have cancer) and H_1 (patient has cancer). The Bayesian formulation of Eq. (3.10) predicts the required probability as

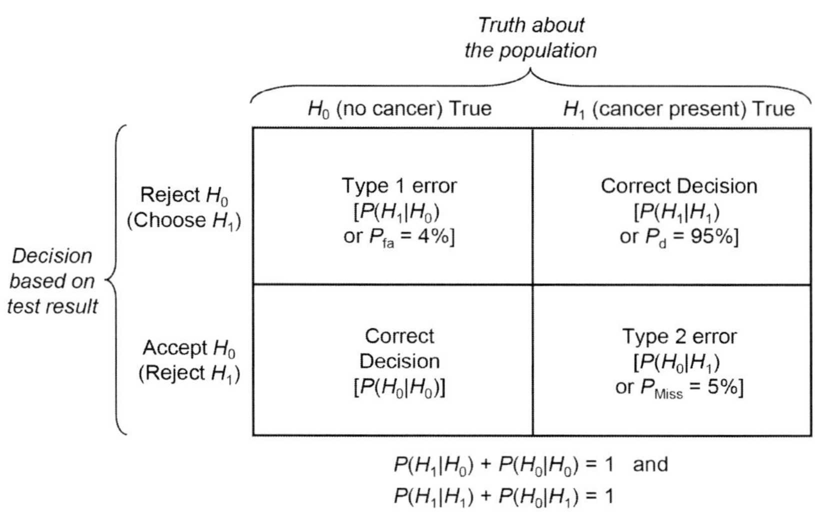

Figure 3.3 Hypotheses and errors in cancer screening example.

$$P(\text{patient has cancer} \mid \text{test positive}) = \frac{P(\text{test positive} \mid \text{cancer})P(\text{cancer})}{P(\text{test positive})}, \tag{3.28}$$

where

$$P(\text{test positive}) = P(\text{test positive}|\text{cancer})P(\text{cancer}) + P(\text{test positive}|\text{no cancer})P(\text{no cancer}). \tag{3.29}$$

The probability P(test positive | no cancer) is the false-alarm probability or Type 1 error. The Type 2 error is the probability of missing the detection of cancer in a patient with the disease.

Upon substituting the statistics from Fig. 3.3 into Eq. (3.28), we find that

$$P(\text{patient has cancer} \mid \text{test positive}) = \frac{(0.95)(0.005)}{(0.95)(0.005) + (0.04)(0.995)} = 0.107 \tag{3.30}$$

or 10.7%.

Intuitively, this result may appear smaller than expected. It asserts that in only 10.7% of the cases in which the test gives a positive result and declares cancer to be present is it actually true that cancer is present. However, the retrospective support given to the cancer hypothesis by the test evidence (through the likelihood function) has increased its degree of belief by approximately a factor of 20, from 5 in 1000 to 107 in 1000.

Further testing is thus required when this type of initial test is administered. The screening test may be said to be reliable because it will detect cancer in 95% of the cases in which cancer is present. However, the critical Type 2 error is 0.05, implying that the test will not diagnose 1 in 20 cancers.

To increase the probability of the patient actually having cancer, given a positive test, and concurrently reduce the Type 2 error requires a test with a greater accuracy. A more-effective method of increasing the *a posteriori* probability is to reduce the false-alarm probability. If, for example, the test accuracy is increased to 99.9% and the false-alarm probability is reduced to 1%, the probability of the patient actually having cancer, given a positive test, is increased to 33.4%. The Type 2 error now implies a missed diagnosis in only 1 out of 1,000 patients. Increasing the test accuracy to 99.99% has a minor effect on the *a posteriori* probability, but it reduces the Type 2 error by another order of magnitude.

In other situations, the Type 1 error may be the more serious error. Such a case occurs if an innocent man was tried for a crime and his freedom relied on the outcome of a certain experiment. If a hypothesis corresponding to his innocence was constructed and was rejected by the experiment, then an innocent man would be convicted and a Type 1 error would result. On the other hand, if the man was guilty and the experiment accepted the hypothesis corresponding to innocence, the guilty man would be freed and a Type 2 error would result. Obviously, the Type 1 error is the more critical one to the person on trial.

Another breast cancer screening test is based on exposing single strands of hair to concentrated X-ray beams and examining the diffraction pattern formed. It is known from test trial data that

$$P(\text{test positive} \mid \text{cancer present}) = P(H_1|H_1) = 0.82 \text{ and} \tag{3.31}$$

$$P(\text{test negative} \mid \text{cancer not present}) = P(H_0|H_0) = 0.77. \tag{3.32}$$

The problem is to find the probability that a patient has cancer given a positive test result when the probability of cancer in the general population is 0.005.

From $P(H_1|H_1) + P(H_0|H_1) = 1$, calculate

$$P(H_0|H_1) = 0.18 = P(\text{test negative} \mid \text{cancer present}). \tag{3.33}$$

From $P(H_1|H_0) + P(H_0|H_0) = 1$, calculate

$$P(H_1|H_0) = 0.23 = P(\text{test positive} \mid \text{cancer not present}), \tag{3.34}$$

where H_1 represents the presence of cancer, and H_0 represents the absence of cancer. The required probability is then found as

$$\left.\begin{aligned} P(\text{patient has cancer}|\text{test pos.}) &= \frac{P(\text{test pos.} \mid \text{cancer})P(\text{cancer})}{P(\text{test pos.})} \\ &= \frac{P(\text{test pos.} \mid \text{cancer})P(\text{cancer})}{P(\text{test pos.} \mid \text{cancer present})P(\text{cancer}) + P(\text{test pos.} \mid \text{no cancer})P(\text{no cancer})} \\ &= \frac{(0.82)(0.005)}{(0.82)(0.005) + (0.23)(0.995)} = 0.0176 \approx 1.8\%. \end{aligned}\right\} \tag{3.35}$$

3.6 Bayesian Inference in Support of Data Fusion

Computing probabilities from subjective information allows the Bayesian inference process to be applied to multisensor fusion since probability density functions are not required. A cautionary note—the output of such a process is only as good as the input *a priori* data.

Figure 3.4 illustrates the Bayesian inference process as applied to the fusion of multisensor data.[9] In this drawing, multiple sensors collect parametric data about an entity whose identity is unknown. For example, typical features used

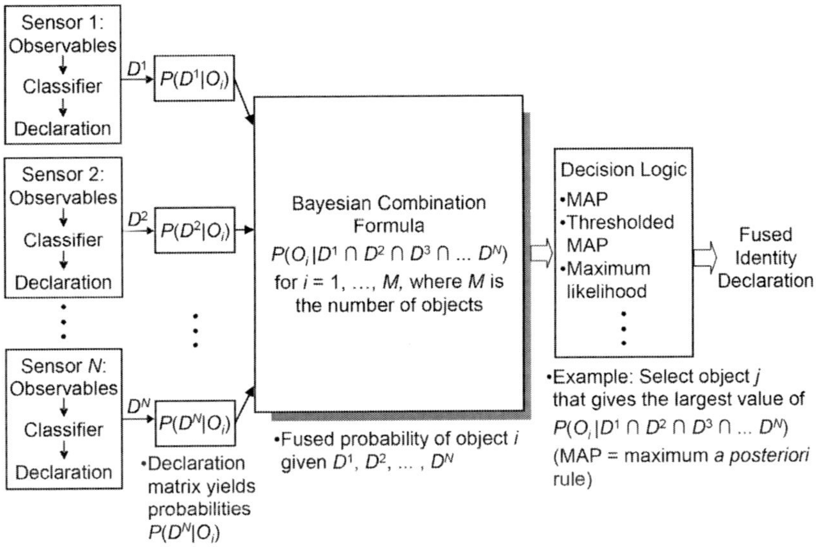

Figure 3.4 Bayesian inference fusion of multisensor data (MAP is maximum *a posteriori* rule) (adapted from Ref. 9).

to identify an object from visible spectrum or infrared video imagery are image-fill criteria where the number of pixels above some threshold is compared to the total number of pixels within the image boundaries, length/width ratio of the image (unnormalized or normalized to area or edge length), parallel and perpendicular line relationships, presence of arcs or circles or conic shapes in the image, central moments, center of gravity, asymmetry measures, and, for infrared imagery, temperature gradients across object boundaries. Frequency-domain signal characteristics that assist in identifying objects are separation between spectral peaks, widths of spectral features, identification of periodic structures in the signal, and number of scattering centers producing a return signal greater than a clutter-adaptive, running-average threshold. Time-domain signal characteristics for object identification include signal amplitudes, pulse repetition intervals, pulse widths, amplitude/width ratios, and rise and fall times of pulses. Other signal features such as radar cross-sections and acoustic signatures may be sensor specific.

Each of the sensors provides an identity declaration D or hypothesis about the object's identity based on the observations and a sensor-specific classification algorithm. The previously established performance characteristics of each sensor's classification algorithm (developed either theoretically or experimentally) provide estimates of the likelihood function, i.e., the probability $P(D|O_i)$ that the sensor will declare the object to be a certain type, given that the object is in fact type i. These declarations are then combined using a generalization of Bayes' theorem [Eq. (3.10)] to produce an updated, joint probability for each entity O_i founded on the multisensor declarations.

Thus, the probability of having observed object i from the set of M objects given declaration (evidence) D^1 from Sensor 1, declaration D^2 from Sensor 2, etc., is given by the Bayesian combination formula

$$P(O_i|D^1 \cap D^2 \cap D^3 \cap \ldots \cap D^N),\ i = 1, \ldots, M. \tag{3.36}$$

By applying a decision logic, a joint declaration of identity can be selected by choosing the object whose joint probability given by Eq. (3.36) is greatest. The choice of the maximum value of Eq. (3.36) is referred to as the maximum *a posteriori* (MAP) probability decision rule. Other decision rules exist as indicated in the figure. The Bayes formulation, therefore, provides a method to combine identity declarations from multiple sensors to obtain a new and hopefully improved joint identity declaration. Required inputs for the Bayes method are the ability to compute or model $P(D|O_i)$ for each sensor and entity, and the *a priori* probabilities that the hypotheses $P(O_i)$ are true. When *a priori* information is lacking concerning the relative likelihood of O_i, the principle of indifference may be invoked in which $P(O_i)$ for all i are initially set equal to one another.

The application of Bayes' theorem is often contrasted in modern probability theory with the application of confidence intervals. While Bayes' theorem provides an inference approach suitable for some data fusion applications, the theory of confidence intervals is better suited when it is desired to assert, with some specified probability, that the true value of a certain parameter (e.g., mean and variance) that characterizes a known distribution is situated between two limits.

The Bayesian approach to recursive updating of the posterior probability occurs by utilizing the previous posteriors as the new values for the prior probabilities. For example, in Eq. (3.37), H_i denotes a hypothesis as before. The vector $\mathbf{E}^N = E^1, E^2, \ldots, E^N$ represents a sequence of data observed from N sources in the past, while E represents a new fact (or new datum). If once we have calculated $P(H_i|\mathbf{E}^N)$ and we can discard past data, then the impact of the new datum E is expressed as[8–10]

$$\begin{aligned} P(H_i|\mathbf{E}^N, E) &= \frac{P(E|\mathbf{E}^N, H_i)P(H_i|\mathbf{E}^N)}{P(E|\mathbf{E}^N)} \\ &= \frac{P(E|\mathbf{E}^N, H_i)P(H_i|\mathbf{E}^N)}{\sum_i [P(E|\mathbf{E}^N, H_i)P(H_i)]}, \end{aligned} \tag{3.37}$$

where

$P(H_i \mid \mathbf{E}^N, E)$ = *a posteriori* probability of H_i for the current period, given the evidence or data $\mathbf{E}^N$, E available at the current period,

$P(E \mid \mathbf{E}^N, H_i)$ = probability of observing evidence E given H_i and the evidence $\mathbf{E}^N$ from past observations (the likelihood function),

$P(H_i \mid \mathbf{E}^N)$ = *a priori* probability of H_i, set equal to the posterior probability calculated using the evidence $\mathbf{E}^N$ from past observations, and
$\sum_i P(E|\mathbf{E}^N, H_i)P(H_i)$ = preposterior or probability of observing evidence $\mathbf{E}^N$ from past observations and the presence of events H_i.

In this updating of the *a posteriori* probabilities, the old belief $P(H_i \mid \mathbf{E}^N)$ assumes the role of the prior probability when computing the new posterior. It completely summarizes past experience. Accordingly, updating of the posterior is accomplished by multiplying the old belief by the likelihood function $P(E \mid \mathbf{E}^N, H_i)$, which equals the probability of the new datum E given the hypothesis and the past observations.

A simplification of Eq. (3.37) occurs under two conditions. The first allows the discarding of past evidence or data once the posterior probability $P(H_i \mid \mathbf{E}^N)$ is calculated. Then the likelihood function is independent of the past data and involves only E and H_i such that

$$P(E|\mathbf{E}^N, H_i) = P(E|H_i), \tag{3.38}$$

where $\mathbf{E}^N$ is the discarded sequence of data observed from N sensors or sources in the past.

The second condition requires the sensors to be conditionally independent. For example, sensors E^1 and E^2 are considered conditionally independent of each other if there is an event H such that

$$p(E^1, E^2|H) = p(E^1|H)p(E^2|H). \tag{3.39}$$

When the N sensors are conditionally independent,

$$P(E^1, E^2, \ldots E^N|H) = \prod_{k=1}^{N} P(E^k)H. \tag{3.40}$$

When both the discarding of past data and conditional independence conditions are met, Bayes' theorem becomes

$$P(H_i|E) = \frac{P(E|H_i)P(H_i)}{P(E)} = \frac{P(E|H_i)P(H_i)}{\sum\limits_i [P(E|H_i)P(H_i)]}, \tag{3.41}$$

which is identical to Eq. (3.10).

3.7 Bayesian Inference Applied to Vehicle Identification

The example in this section describes how to apply multisensor data to recursively update the posterior probability when multiple hypotheses exist. It follows a method in Pearl[8] that selects the most likely hypothesis based on the output states of several sensors. Another example is found in Section 2 of

Durrant-Whyte,[2] where observations from two sensors return values of likelihood functions for a three-object identification problem.

Suppose that several hypotheses $\mathbf{H} = \{H_1, H_2, H_3, H_4\}$ exist, where each represents the occurrence of one of four possible vehicles on a highway, such as

H_1 = passenger vehicle on a highway,
H_2 = bus on a highway,
H_3 = 18-wheeler on a highway, and
H_4 = motorcycle on a highway.

Assume that the evidence variable $\mathbf{E}^k$ produced by a sensor has one of several output states in response to the detection of a vehicle. For example, when a multispectral sensor is used, three types of outputs may be available as represented by

E_1^k = evidence from detected emission in radiance spectral band 1,
E_2^k = evidence from detected emission in radiance spectral band 2, and
E_3^k = evidence from detected emission in radiance spectral band 3.

The causal relations between $\mathbf{H}$ and $\mathbf{E}^k$ are displayed in a $q \times r$ matrix $\mathbf{M}^k$, where q is the number of hypotheses under consideration, and r is the number of output states or output values of the sensor. The $(i, j)^{\text{th}}$ matrix element of $\mathbf{M}^k$ represents

$$M_{ij}^k = P(E_j^k | H_i). \tag{3.42}$$

For instance, the sensitivity of the kth sensor having $r = 3$ output states to $\mathbf{H}$ containing $q = 4$ hypotheses is represented by the 4×3 evidence matrix in Table 3.3. On the basis of the given evidence, the overall belief in the ith hypothesis H_i is [from Eq. (3.10) or (3.41)]

$$P(H_i | E_1, \ldots, E_r) = \alpha P(E_1, \ldots, E_r | H_i) P(H_i), \tag{3.43}$$

where α is a normalizing factor computed by requiring Eq. (3.41) to sum to unity over i. Accordingly, α equals the inverse of the sum of the elements of $P(E_1, \ldots, E_r | H_i) P(H_i)$. In contrast, the likelihood functions $P(E^k | H_i)$ are not required to sum to unity over i.

Table 3.3 Likelihood functions $P(E^k|H_i)$ corresponding to evidence produced by the kth sensor with three output states in support of four hypotheses.

	E_1^k: Detection of emission in spectral band 1	E_2^k: Detection of emission in spectral band 2	E_3^k: Detection of emission in spectral band 3
H_1	0.35	0.40	0.10
H_2	0.26	0.50	0.44
H_3	0.35	0.10	0.40
H_4	0.70	0	0

When a sensor's response is conditionally independent, i.e., each $P(E_j|H_i)$ is independent of that of the other sensors, Eq. (3.40) can be applied to give

$$P(H_i|E_1, \ldots, E_r) = \alpha P(H_i) \prod_{k=1}^{N} P(E^k|H_i). \tag{3.44}$$

A likelihood vector $\boldsymbol{\lambda}^k$ is defined to describe the evidence E^k produced by each sensor as

$$\boldsymbol{\lambda}^k = (\lambda_1^k, \lambda_2^k, \ldots, \lambda_q^k), \tag{3.45}$$

where

$$\lambda_i^k = P(E^k|H_i). \tag{3.46}$$

Equation (3.44) can be evaluated using a two-step vector-product process as follows:

1. The individual likelihood vectors from each sensor are multiplied together, term by term, to obtain an overall likelihood vector $\boldsymbol{\Lambda} = \boldsymbol{\lambda}_1, \ldots, \boldsymbol{\lambda}_q$ given by

$$\Lambda_i = \prod_{k=1}^{N} P(E^k|H_i). \tag{3.47}$$

2. The overall belief vector $P(H_i|E^1, \ldots, E^N)$ is computed from the product

$$P(H_i|E^1, \ldots, E^N) = \alpha P(H_i)\Lambda_i. \tag{3.48}$$

Only estimates for the relative magnitudes of the conditional probabilities in Eq. (3.46) are required. Absolute magnitudes do not affect the outcome because α can be found later from the requirement

$$\sum_i P(H_i|E^1, \ldots, E^N) = 1. \tag{3.49}$$

To model the behavior of a multisensor system, let us assume that two sensors are deployed, each having the identical evidence matrix shown in Table 3.3. Furthermore, the prior probabilities for the hypotheses $\mathbf{H} = \{H_1, H_2, H_3, H_4\}$ are assigned as

$$P(H_i) = (0.42,\ 0.25,\ 0.28,\ 0.05), \tag{3.50}$$

where Eq. (3.11) is satisfied by this distribution of prior probabilities.

If Sensor 1 detects emission in spectral band 3 and Sensor 2 detects emission in spectral band 1, the elements of the likelihood vector are

$$\boldsymbol{\lambda}^1 = P(E^1|H_i) = (0.10,\ 0.44,\ 0.40,\ 0) \tag{3.51}$$

and

$$\boldsymbol{\lambda}^2 = P(E^2|H_i) = (0.35,\ 0.26,\ 0.35,\ 0.70). \tag{3.52}$$

Therefore, the overall likelihood vector is

$$\boldsymbol{\Lambda} = \boldsymbol{\lambda}^1\boldsymbol{\lambda}^2 = (0.035,\ 0.1144,\ 0.140,\ 0), \tag{3.53}$$

and from Eq. (3.48),

$$\begin{aligned} P(H_i|E^1,\ E^2) &= \alpha P(H_i)\Lambda \\ &= \alpha(0.42,\ 0.25,\ 0.28,\ 0.05)\cdot(0.035,\ 0.1144,\ 0.140,\ 0) \\ &= \alpha(0.0147,\ 0.0286,\ 0.0392,\ 0) = (0.178,\ 0.347,\ 0.475,\ 0), \end{aligned} \tag{3.54}$$

where α is found from the requirement of Eq. (3.49) as the inverse of the sum of $0.0147 + 0.0286 + 0.0392 + 0$, which equals 12.1212.

From Eq. (3.54), we conclude that the probability of a large vehicle on the highway, H_2 or H_3, is $0.347 + 0.475 = 0.822$ or 82.2%, and the probability of a passenger vehicle being present is 17.8%. The combined probability for some vehicle being present is 100%.

Updating of the posterior belief does not have to be delayed until all of the evidence is collected but can be implemented incrementally. For example, if it is first observed that Sensor 1 detects emission in spectral band 3, the belief in $\mathbf{H}$ becomes

$$\begin{aligned} P(H_i|E^1) &= \alpha(0.42,\ 0.25,\ 0.28,\ 0.05)\cdot(0.10,\ 0.44,\ 0.40,\ 0) \\ &= \alpha(0.042,\ 0.110,\ 0.112,\ 0) = (0.159,\ 0.417,\ 0.424,\ 0), \end{aligned} \tag{3.55}$$

with $\alpha = 3.7879$.

These values of the posterior probabilities are now utilized as the new values of the prior probabilities when the next datum arrives, namely, evidence from Sensor 2, which detects emission in spectral band 1. Figure 3.5 demonstrates the recursive nature of the posterior probability updating process.

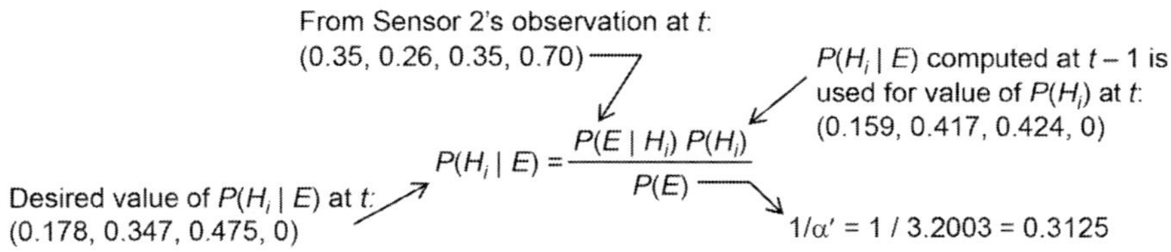

Figure 3.5 Recursive updating of the posterior probability for the example in Section 3.7.

Upon incorporating the evidence from Sensor 2, the posterior probability becomes

$$\begin{aligned}P(H_i|E^1, E^2) &= \alpha'\lambda_i^2 \cdot P(H_i|E^1)\\ &= \alpha'(0.35, 0.26, 0.35, 0.70) \cdot (0.1591, 0.4167, 0.4242, 0)\\ &= \alpha'(0.0557, 0.1083, 0.1485, 0) = (0.178, 0.347, 0.475, 0),\end{aligned} \tag{3.56}$$

where $\alpha' = 3.2003$. This is the same result given by Eq. (3.54) for $P(H_i|E^1, E^2)$.

Thus, the evidence from Sensor 2 lowers the probability of a large vehicle being present slightly from 84.1% to 82.2% but increases the probability of a passenger vehicle being present by the same amount from 15.9% to 17.8%. The result specified by Eq. (3.54) or Eq. (3.56) is unaffected by which sensor's evidence arrives first and is subsequently used to update the priors for incorporation of the evidence from the next datum.

The incident detection application of Bayesian inference presented below utilizes information from roadway sensors, cell phones, and commercial truck radio transmissions. In this case, the likelihood functions are not constant but are dependent on weather and lighting conditions among other factors.

3.8 Bayesian Inference Applied to Freeway Incident Detection Using Multiple-Source Data

Incident detection may be enhanced by fusing data from more than one information source if each source produces a signature or data generated by conditionally independent sources. Suppose that a scenario exists where traffic flow data and incident reports are available from roadway sensors, cellular telephone calls from travelers, and radio reports from commercial truck drivers.[4,10] Each of these data sources has limitations that prevent using any one of them alone for reliable incident detection. For instance, sensor spacing, delay in acquiring sufficient data to detect an incident, or false-alarm history is not adequate to detect or confirm an incident with a sufficiently high probability (>80%) in a timely manner from unaccompanied roadway sensor data. The cellular phone calls are known to contain inaccurate incident location data, and the radio reports are too infrequent to confirm the incidents by themselves.

Using historical data, traffic management personnel serving the affected area have constructed *a priori* probabilities for the likelihood that roadway sensor data are reporting a true incident based on the length of time that lane occupancy (i.e., percent of selected time interval that vehicles are detected in the detection area of a sensor) and traffic volume are above preset thresholds, and that speed is below some other threshold. *A priori* probabilities also are assumed to be available to describe the accuracy of the cellular telephone and radio incident reports as a function of the number of calls and the variance of the reported incident locations.

3.8.1 Problem development

We wish to apply Bayesian inference to compute the *a posteriori* probabilities that the detected incident event belongs to one of three types:

H_1 = one or more vehicles is on the right shoulder of the highway,
H_2 = traffic in right-most lane is slower than normal, or
H_3 = traffic is flowing normally in all lanes.

The Bayesian approach to data fusion updates the posterior probability as illustrated in the influence diagram of Fig. 3.6. The probability of the road sensors (RS) reporting data conditioned on event type j is given by

$$\begin{aligned} P_{\mathrm{RS}}(\mathrm{data}|\mathrm{Event}_j) &= P_{\mathrm{RS}}(\mathrm{data}|H_1)P(H_1|\mathrm{Event}_j) \\ &+ P_{\mathrm{RS}}(\mathrm{data}|H_2)P(H_2|\mathrm{Event}_j) + P_{\mathrm{RS}}(\mathrm{data}|H_3)P(H_3|\mathrm{Event}_j), \end{aligned} \tag{3.57}$$

the probability of the cellular telephone (CT) calls reporting data conditioned on event type j by

$$\begin{aligned} P_{\mathrm{CT}}(\mathrm{data}|\mathrm{Event}_j) &= P_{\mathrm{CT}}(\mathrm{data}|H_1)P(H_1|\mathrm{Event}_j) \\ &+ P_{\mathrm{CT}}(\mathrm{data}|H_2)P(H_2|\mathrm{Event}_j) + P_{\mathrm{CT}}(\mathrm{data}|H_3)P(H_3|\mathrm{Event}_j), \end{aligned} \tag{3.58}$$

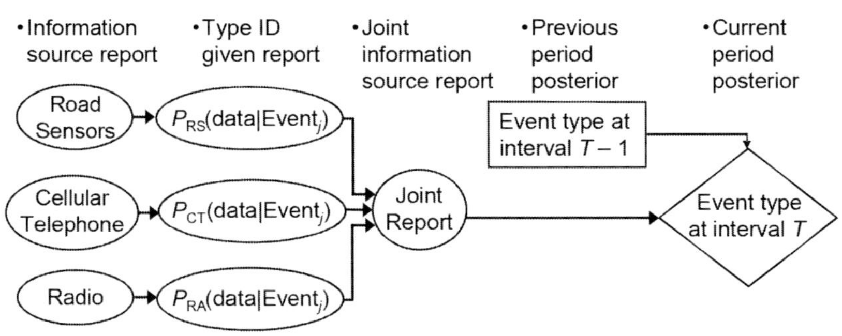

Figure 3.6 Influence diagram for freeway event detection using data from three conditionally independent sources.

and the probability of truck driver radios (RA) reporting data conditioned on event type j by

$$\begin{aligned} P_{\mathrm{RA}}(\mathrm{data}|\mathrm{Event}_j) &= P_{\mathrm{RA}}(\mathrm{data}|H_1)P(H_1|\mathrm{Event}_j)+ \\ &P_{\mathrm{RA}}(\mathrm{data}|H_2)P(H_2|\mathrm{Event}_j) + P_{\mathrm{RA}}(\mathrm{data}|H_3)P(H_3|\mathrm{Event}_j), \end{aligned} \tag{3.59}$$

where Event_j is one of the three events H_1, H_2, H_3. The set of arrows from "Information source report" to "Type ID given report" in Fig. 3.6 represents the probability calculations defined by Eqs. (3.57) through (3.59).

The values of the likelihood functions for the roadway sensors $P_{\mathrm{RS}}(\mathrm{data}|H_1)$, $P_{\mathrm{RS}}(\mathrm{data}|H_2)$, and $P_{\mathrm{RS}}(\mathrm{data}|H_3)$; cellular telephone $P_{\mathrm{CT}}(\mathrm{data}|H_1)$, $P_{\mathrm{CT}}(\mathrm{data}|H_2)$, and $P_{\mathrm{CT}}(\mathrm{data}|H_3)$; and the radio $P_{\mathrm{RA}}(\mathrm{data}|H_1)$, $P_{\mathrm{RA}}(\mathrm{data}|H_2)$, and $P_{\mathrm{RA}}(\mathrm{data}|H_3)$ are found through *a priori* measurements, and data collection and analysis. Lane occupancy, traffic volume, and speed measured by the road sensors are compared with predetermined or real-time calculated thresholds, depending on the incident detection algorithm, distance between sensors, and the data-reporting interval characteristics of declaring an event of type j. Thus, offline analysis of the values and duration of real-time data determines the value of the likelihood function that expresses the probability that the data represent hypothesis H_i.

Cellular telephone data are the number of calls that report the same event and the variance of the reported event location. The value of the likelihood function $P_{\mathrm{CT}}(\mathrm{data}|\mathrm{Event}_j)$ equals the probability of receiving a predetermined number of calls with a predetermined event location variance, given the event is of type j. These probabilities are found from historical data collected as a function of event type, number of lanes affected, road configuration, traffic volume, weather, time of day, day of week, season, lighting, etc. Similar data based on trucker behavior are used to define the likelihood functions for the radio reports.

Quantitative values for the *a priori* probabilities $P(H_1|\mathrm{Event}_j)$, $P(H_2|\mathrm{Event}_j)$, and $P(H_3|\mathrm{Event}_j)$ are determined from offline analysis of the types and numbers of events in the monitored area.

Next, the joint information source report shown in Fig. 3.6 is computed for a given time interval as the product of Eqs. (3.57) through (3.59) because the information sources are presumed to generate data from conditionally independent sources. Thus, the joint information source report is

$$P(\mathrm{data}|\mathrm{Event}_j) = \prod_k P^k(\mathrm{data}|\mathrm{Event}_j), \tag{3.60}$$

where k is the information source index, here equal to 1, 2 and 3 for road sensor, cellular telephone, and radio, respectively.

Finally, Bayes' theorem is applied to calculate the current period *a posteriori* probability $P(\mathrm{Event}_j|\mathrm{data})$ that the detected event is of type j based on the values of the posterior probabilities evaluated during the previous period and the joint information source report. Accordingly,

$$P(\text{Event}_j|\text{data}) = \frac{P(\text{data}|\text{Event}_j)P(\text{Event}_j)}{P(\text{data})}, \tag{3.61}$$

where

$$P(\text{Event}_j) = \text{value of } P(\text{Event}_j|\text{data}) \text{ during the previous period, and} \tag{3.62}$$

$$P(\text{data}) = \sum_j P(\text{data}|\text{Event}_j)P(\text{Event}_j) \tag{3.63}$$

is the preposterior or probability of observing the data collected during the previous period given that events denoted by Event_j are present. Larger values of $P(\text{data})$ imply that the previous period values are more predictive of the situation as it evolves; i.e., the change in $P(\text{Event}_j|\text{data})$ from the previous period to the current period is smaller. When the information sources do not report an event type for the current time interval, updating is not performed and the values of $P(\text{Event}_j|\text{data})$ for the current interval are set equal to those from the previous period.

If the primary task is to detect abnormal traffic flow or an incident, the first and second terms in Eqs. (3.57) through (3.59) can be combined into a single declaration INCIDENT. The problem is further simplified by choosing $\text{Event}_1 = \text{INCIDENT}$ and $\text{Event}_2 = \overline{\text{INCIDENT}}$, where the bar denotes negation. Therefore, the required probabilities are only dependent on $P(\text{data}|\text{Event}_1)$ since

$$P(\text{data}|\text{Event}_2) = 1 - P(\text{data}|\text{Event}_1), \text{ and} \tag{3.64}$$

$$P(\text{Event}_2) = 1 - P(\text{Event}_1). \tag{3.65}$$

Returning to the three-hypothesis problem, an initial value for $P(\text{Event}_j)$ and lower and upper bounds inside the interval (0, 1) for admissible values of $P(\text{Event}_j|\text{data})$ are needed to evaluate Eq. (3.61). When information concerning the initial values of $P(H_1)$, $P(H_2)$, and $P(H_3)$ is lacking, the initial values are set equal to one another with the value of 1/3 (i.e., the insufficient reason principle is applied). The boundaries for $P(H_1|\text{data})$, $P(H_2|\text{data})$, and $P(H_3|\text{data})$ are limited to (0.01, 0.99) to prevent the process that computes the *a posteriori* probability from terminating prematurely at the limiting endpoint values of 0 and 1.

The posterior probabilities for events H_1, H_2, and H_3 are computed from the posterior probabilities for event type and the scenario-defined values for $P(H_1|\text{Event}_j)$, etc., as

$$P(H_1|\text{data}) = \sum_j [P(\text{Event}_j|\text{data})P(H_1|\text{Event}_j), \tag{3.66}$$

$$P(H_2|\text{data}) = \sum_j [P(\text{Event}_j|\text{data})P(H_2|\text{Event}_j), \text{ and} \tag{3.67}$$

$$P(H_3|\text{data}) = \sum_j [P(\text{Event}_j|\text{data})P(H_3|\text{Event}_j). \tag{3.68}$$

Thus, the probability of determining whether an incident has occurred is the sum of individual probabilities that are dependent on the identification of various features. The term $P(H_1|\text{Event}_j)$ expresses the *a priori* probability of finding event H_1 conditioned on event type j being present. Similar interpretations for $P(H_2|\text{Event}_j)$ and $P(H_3|\text{Event}_j)$ apply. Practical applications require the data fusion algorithms to be robust in their ability to identify the obvious events and those that are unexpected. It is also beneficial to have information sources at your disposal that can assist in the detection and identification of more than one type of event.

3.8.2 Numerical example

Assume that the likelihood functions $P(\text{data}|H_i)$ are specified by the entries in Tables 3.4 through 3.6 for the road sensors, cellular telephone calls, and radio reports, respectively, and are based, in general, on the considerations

Table 3.4 Road sensor likelihood functions for the three-hypothesis freeway incident detection problem.

	E^{RS}: Probability of data representing H_i
H_1	0.15
H_2	0.70
H_3	0.85

Table 3.5 Cellular telephone call likelihood functions for the three-hypothesis freeway incident detection problem.

	E^{CT}: Probability of data representing H_i in good weather	E^{CT}: Probability of data representing H_i in inclement weather	E^{CT}: Probability of data representing H_i in darkness or poor lighting conditions
H_1	0.46	0.35	0.25
H_2	0.60	0.43	0.35
H_3	0.90	0.75	0.65

Table 3.6 Radio report likelihood functions for the three-hypothesis freeway incident detection problem.

	E^{RA}: Probability of data representing H_i in good weather	E^{RA}: Probability of data representing H_i in inclement weather	E^{RA}: Probability of data representing H_i in darkness or poor lighting conditions
H_1	0.60	0.50	0.45
H_2	0.85	0.75	0.65
H_3	0.98	0.85	0.75

discussed following Eq. (3.59). Only one set of road sensor likelihood functions is utilized as the parameters on which the effectiveness of the sensors in reporting incidents, namely, the incident detection algorithm, the distance between sensors, and the data-reporting interval, are assumed to be known and constant. The parameters depicted for the likelihood functions of the cellular telephone calls and radio reports are representative of those upon which these likelihood functions may depend. Further assume that the prior probabilities are known and given by

$$P(H_i) = (0.5,\ 0.3,\ 0.2). \tag{3.69}$$

Under inclement weather conditions, the overall likelihood vector that represents the combined evidence from the three sensor types is

$$\begin{aligned}\boldsymbol{\Lambda} = \lambda^1\lambda^2\lambda^3 &= (0.15,\ 0.70,\ 0.85)\cdot(0.35,\ 0.43,\ 0.75)\cdot(0.50,\ 0.75,\ 0.85)\\ &= (0.02625,\ 0.2258,\ 0.5419)\end{aligned} \tag{3.70}$$

from application of Eqs. (3.46) and (3.47).

The posterior probability becomes [from Eq. (3.48)]

$$\begin{aligned}P(H_i|E^{RS},\ E^{CT},\ E^{RA}) &= \alpha P(H_i)\Lambda_i\\ &= \alpha(0.5,\ 0.3,\ 0.2)\cdot(0.02625,\ 0.2258,\ 0.5419)\\ &= \alpha(0.0131,\ 0.0677,\ 0.1083),\end{aligned} \tag{3.71}$$

where $\alpha = 1/(0.0131 + 0.0677 + 0.1083) = 5.2882$. Thus,

$$P(H_i|E^{RS},\ E^{CT},\ E^{RA}) = (0.0693,\ 0.3580,\ 0.5727). \tag{3.72}$$

The output of the data fusion process in this example is to declare H_3 the most likely hypothesis, namely, that traffic is flowing normally in all lanes.

3.9 Bayesian Inference Applied to Truck Classification

Hernandez, Tok, and Ritchie[11] investigated the integration of WIM system attributes and high-resolution waveform signatures produced by ILDs to classify trucks into 14 FHWA classes. This topic was discussed in Section 2.13.8 as an example of a Bayesian data fusion application to traffic monitoring. Here we take a more detailed look at the Bayesian process that combines the outputs of the five classifiers in the multiple-classifier system (MCS).

Table 3.7 describes the operating principles of the naïve Bayes (NB), decision tree (DT), support vector machine (SVM), multilayer feedforward (MLFF) neural network, and probabilistic neural network (PNN) classifiers that comprise the MCS. This set of classifiers has the ability to process noisy data, be robust in the presence of class imbalance, and not overlap with the expertise of other classifiers in the ensemble. The MCS was able to outperform the individual classifiers although there was no *a priori* guarantee of this benefit. The MCS also reduced statistical, representational, and computational risks while increasing the performance and generalization capability of the system.

3.9.1 MCS architecture

The MCS architecture depicted in Fig. 3.7 consists of the set of classifiers, likelihood functions that incorporate evidence from observations, and a mechanism to apply Bayes' theorem to combine the individual classifier predictions into a single forecast under the assumption that the classifiers are conditionally independent given a class label. The literature indicates that naïve Bayes methods produce optimal results with respect to misclassification rate even when the classifier independence assumption is violated by a wide margin.[12] Prior to field testing the MCS, the individual classifiers were trained, their performance was optimized through rule formation and identification of decision boundaries that differed for each classifier, and estimates were made of the confusion matrices that approximate the likelihood functions.

Table 3.7 Classifier types in the multiple-classifier system.

Classifier	Operating Principle
Naïve Bayes	Statistical classifier—follows from Bayes' theorem of conditional probabilities and assumes conditional independence of features.
Decision tree	Rule-forming classifier—partitions the feature space into cuboid regions.
Support vector machine	Establishes a set of high-dimensional vectors from training data that outline a separating hyperplane fit to a particular kernel function.
Multilayer feedforward neural network	Nonparametric—network of input, hidden, and output layers of perceptrons using an error-correcting backpropagation training algorithm.
Probabilistic neural network	Radial basis function network that uses statistical distributions, i.e., Gaussian kernel functions, as activation functions.

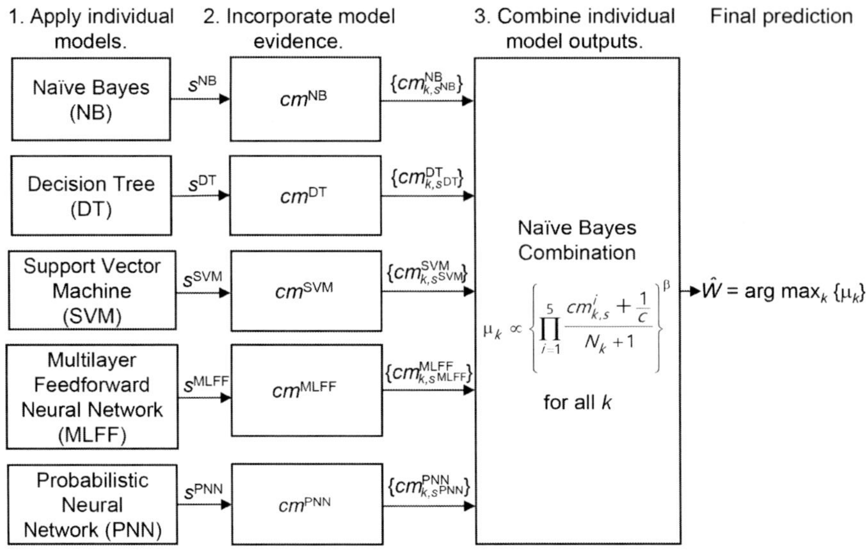

Figure 3.7 MCS architecture application to truck classification (adapted from Ref. 11).

The architecture incorporates the classifications s^i from each individual classifier, utilizes evidence from stored confusion matrices to determine conditional probabilities for each truck class k, and combines the classifier outputs using a form of Bayes' theorem to compute the value of support μ_k for each class to obtain the final truck class prediction $\hat{W}$.

Bayes' theorem gives the posterior probability $P(w_k|s)$ for the predicted class w_k as

$$P(w_k|s) = \frac{P(w_k)P(s|w_k)}{P(s)} = \frac{P(w_k)\prod_{i=1}^{5} P(s^i|w_k)}{P(s)}, \tag{3.73}$$

where $k = 1, \ldots, c$ represents one of the c truck classes, $i = 1, \ldots, 5$ denotes one of the five classifiers in the MCS, $P(w_k)$ are the prior probabilities for the presence of a particular truck class k, and $P(s^i|w_k)$ are the likelihood functions for each truck class s and classifier.

The likelihood functions for each classifier are represented by the elements $cm^i_{k,s}$ of a $c \times c$ cross-classification matrix (also called a confusion matrix), of which there are five in this study. The cross-classification matrix is populated by subjecting a trained classifier to an independent validation data set such that the diagonal elements of the matrix reflect the number of correctly classified trucks, while the off-diagonal elements reflect the number of misclassified trucks. The likelihood functions $P(s^i|w_k)$ correspond to $cm^i_{k,s}/N_k$, where N_k is the number of validation samples in class k. This procedure is similar to the third technique discussed in Section 4.7 for constructing probability mass matrices used in Dempster–Shafer evidential reasoning. The prior probabilities $P(w_k)$ are estimated as N_k/N, where N is the total number of samples in the validation data set.

Instead of assigning a weight to a classifier output, a value of support is calculated for each truck class k. The values of support μ_k are related to Eq. (3.73) with the following modifications:

- $P(s)$ is ignored since it does not depend on w_k and may be treated as a normalizing factor.
- The likelihood functions are characterized by $cm^i_{k,s}/N_k$.
- Zero values of $cm^i_{k,s}$ are negated by including the term $1/c$ in the numerator since a value of zero would force the equation to zero regardless of other classifier estimates.
- Prior probabilities are not included.

With these alterations, the value of support becomes

$$\mu_k \propto \left\{ \prod_{i=1}^{5} \frac{cm^i_{k,s} + \frac{1}{c}}{N_k + 1} \right\}^{\beta}. \tag{3.74}$$

The tunable constant β in Eq. (3.74) typically is a value between 0.5 and 1. Hernandez, Tok, and Ritchie[11] determined that its value had little effect on the overall results and therefore equated it to 0.5 in all models. Finally, the class $\hat{W}$ with the largest value of support is assigned as the final prediction of the ensemble of classifiers as

$$\hat{W} = \arg\max_k \{\mu_k\}. \tag{3.75}$$

3.9.2 MCS operation

Figure 3.8 illustrates the sequence of operations used by the MCS to obtain the truck classification. The first stage analyzes the WIM measurements (axle count, axle spacing, and gross vehicle weight) to classify the trucks into the FHWA axle-based 14 classes used in California.

The second stage introduces the inductive signature features into a MLFF neural network having two hidden layers and 15 nodes per layer. The neural network detects the presence of small, lightweight trailers in FHWA Class 5 trucks that are not identified through analysis of the WIM data and further distinguishes axle configurations in FHWA Class 8 multi-unit trucks. FHWA Class 8 multi-unit trucks had a variety of axle configurations, including three- or four-axle semi-tractor trailers, two-axle single-unit trucks or small trucks with trailers, and three-axle trucks with lift axles (misclassified from FHWA Class 7). The MLFF neural network assigns approximately 71% of three- or four-axle semi-tractor trailers to FHWA Class 8 and thus is successful in identifying this truck category.

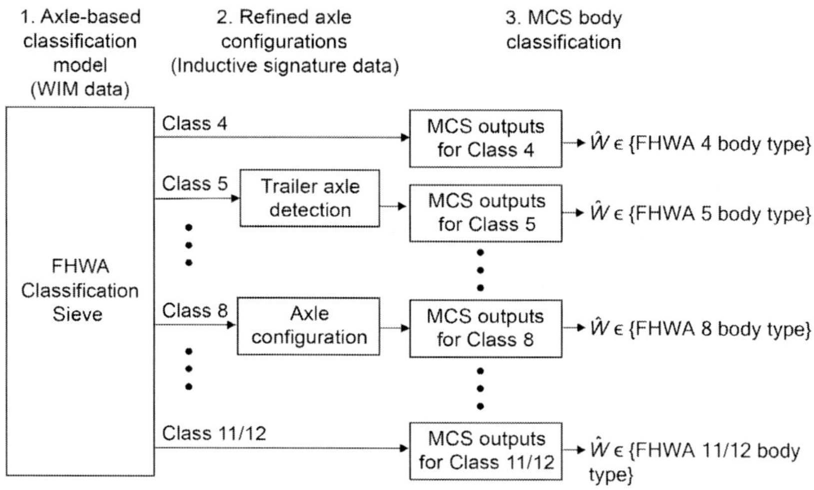

Figure 3.8 MCS stages of operation (adapted from Ref. 11).

The third stage combines the WIM measurements and inductive signature features using the MCS to forecast the final truck body classification. Nine separate body class models were developed, each corresponding to an FHWA axle-based category. Data from trucks without trailers were input into the MCS for FHWA Class 5. Exceptions to this scheme were made for multi-trailer units in FHWA Classes 11 and 12 and for FHWA Class 13 trucks. FHWA Classes 11 and 12 were combined into one body class model because they share the same trailer body types, although they differ in axle count and configuration. Multi-trailer configurations consist of two trailers of the same body type. Therefore, the body class model outputs a single prediction that represents all of the trailers attached to a power unit. Limited samples and a wide diversity in body types of FHWA Class 13 trucks, typically specialized equipment movers or other unique body types, result in no body classification model being developed for this class.

3.9.3 Data collection and conclusions

In California, truck body types vary by location. Therefore, to capture the full diversity of the truck population, data were collected at four disparate locations (Redding, Willows, Fresno, and Irvine) across the state, spanning metropolitan, agricultural, and forested regions, and covered a diverse set of land uses and industries. Ten separate body classification models were developed from an extensive data set of 18,967 truck records, distinguishing a total of 23 single-unit truck and 31 single- and semi-trailer body configurations. The classification results were summarized by a class-specific correct classification rate (CCR) calculated as the ratio of correctly classified vehicles in a class divided by the total number of vehicles in that class. Across all axle configuration groups, the

MCS approach had higher overall CCR than any single-classifier model. While several individual classifiers demonstrated adequate performance at predicting certain classes, there was a consistent tradeoff with low performance of the same classifier on another class or set of truck classes. Using the MCS approach, each of the ten FHWA classes (Classes 4 through 12 and 14) achieved an overall CCR above 80%. These results represent a significant improvement over the performance of the best individual classifier.

The model for FHWA Class 4 distinguished buses from vans and platforms that were misclassified by WIM controllers into this class. For two-axle single-unit trucks within FHWA Class 5, body configurations are extremely heterogeneous, and axle measurements and weights overlap significantly. However, the model effectively separated buses, passenger vans, utility and pickup trucks, bobtails (tractor drive units), and enclosed vans. Models for FHWA Classes 6 and 7 differentiated several industry-specific categories including concrete, dump, garbage, dumpster transport, and vans.

Five-axle combination trucks within FHWA Class 9a possessed the widest diversity of trailer body types. Enclosed vans formed the majority, while several unique and industry-specific classes such as logging and intermodal containers were in the minority. While even the weakest model was able to predict the majority class with at least 60% accuracy, the MCS method was highly beneficial for minority classes, outperforming even the best single classifier by 6.3%. The model also distinguished refrigerated from nonrefrigerated vans as well as refrigerated from nonrefrigerated 40-ft intermodal containers. This level of distinction opens new possibilities for advanced tracking of perishable commodities, especially those that are associated with intermodal facilities such as seaports. Other commodity-specific body types such as automobile transport, logging, and livestock trailers had CCRs above 85%. Unfortunately, the model was not able to accurately distinguish 53-ft intermodal containers from enclosed vans due to overlap in signature and axle configuration features.

3.10 Causal Bayesian Networks

Causal Bayesian networks find relevance in vehicle safety and traffic management applications such as vehicle motion estimation related to a driver's intention,[13] fusion of sensor (hard data) and human (soft data) information,[14] traffic modeling and performance prediction when system parameters vary,[15] fault diagnosis of braking systems,[16] and vehicle trajectory prediction and criticality assessment.[17]

Their value in these and other circumstances comes from the ability to model the influence of a cause on the effect it produces with a joint probability distribution, which leads to the calculation of the posterior probability using Bayes' theorem. The remainder of this section introduces the theory of causal Bayesian networks and provides an example of their utilization for vehicle trajectory prediction.

3.10.1 Directed acyclic graphs

Causal Bayesian networks are directed graphs that simulate links between nodes that represent a cause and an effect. The nodes are variables that represent an effect or action, observed value, default value, or a value calculated from information in a node that precedes it. The relationship between nodes is expressed by their connections. Hence, if two nodes are connected by a direct line, the strength of which is represented by the conditional probability $P(N|M)$, the value of node M is said to directly influence the value of node N. The discussion below concerning graph theory is based on material from Greenland and Pearl.[18]

3.10.1.1 Underlying theory

Causal system models can be represented by graphs or path diagrams. For instance, Fig. 3.9 depicts a graph model corresponding to any system of five equations that embody the following assumptions:[18]

1. Independence of A and B.
2. Direct dependence of C on A and B.
3. Direct dependence of E on A and C.
4. Direct dependence of F on C.
5. Direct dependence of D on B, C, and E.

The dependence of a node (or vertex) in a graph on another node is shown by the connectivity of the two nodes. The variables represented by the nodes are joined by arcs, edges, or links. If the arc is an arrow, the tail of an arrow signifies the starting variable, called the *parent*, and the head or ending variable is the *child*. According to Greenland and Pearl, an arrow in a causal diagram "represents a *direct effect* of the parent on the child, although this effect is direct only relative to a certain level of abstraction in that the graph omits any variables that might mediate the effect."[18]

A variable without a parent (such as A and B in Fig. 3.9) is exogenous or external, or a root or source node, and is determined solely by forces outside of the graph. Otherwise, it is endogenous or internal. A variable with no children (such as D) is a sink or terminal node. The set of all parents of a

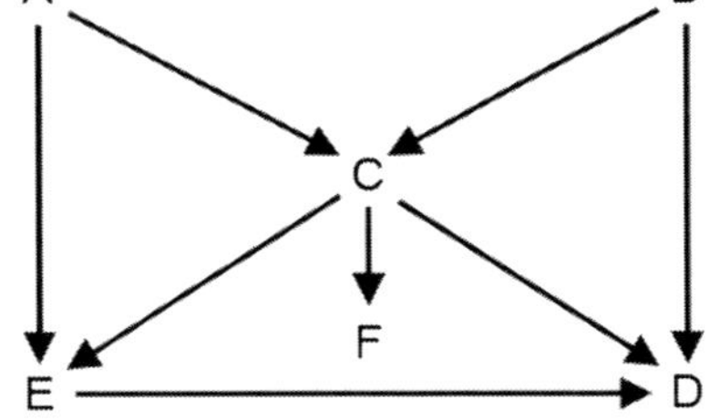

Figure 3.9 Causal Bayesian network representing the graph model in Section 3.10.1.1.

variable X (all variables at the tail of an arrow pointing into X) is symbolized by $pa[X]$. In Fig. 3.9, $pa[D] = \{B, C, E\}$.

A path or chain is a sequence of adjacent arcs. A *directed path* is a path traced out entirely along arrows tail-to-head. If there is a directed path from X to Y, X is an *ancestor* of Y, and Y is a *descendant* of X. In causal diagrams, directed paths represent causal pathways from the starting variable to the ending variable. Thus, a variable can be described as a cause of its descendants and an effect of its ancestors. In a *directed graph* the only arcs are arrows.

An *acyclic graph* is one in which there are no feedback loops (directed paths from a variable back to itself). Hence, Greenland and Pearl define a *directed acyclic graph* (DAG) as "a graph with only arrows for arcs and no feedback loops, that is, no variable is its own ancestor or its own descendant. A DAG represents a complete causal structure in that all sources of dependence are explained by causal links."[18]

A variable *intercepts* or *mediates* a path if it is in the path (but not at the ends). Similarly, a set of variables S intercepts a path if it contains any variable intercepting the path. Variables that intercept directed paths are referred to as *intermediates* on the pathway.

A variable is a *collider* on a path if the path enters and leaves the variable via arrowheads. Being a collider is relative to a path. For instance, in Fig. 3.9, C is a collider on the path A → C ← B → D and a noncollider on the path A → C → D. Nonetheless, it is common to refer to a variable as a collider if it is a collider along any path, i.e., if it has more than one parent. A path is open or unblocked at noncolliders and closed or blocked at colliders. Therefore, a path with no collider, such as E ← C ← B → D, is open or active, while a path with a collider, such as E ← A → C ← B → D, is closed or inactive.

Two variables or sets of variables in the graph are *separated* if there is no open path between them. Independencies arising from separation impose some of the most important constraints on variables in a graphical model. As an example, consider the absence of an open path from A to B in Fig. 3.9 that constrains A and B to be marginally independent, inferring that they are independent if no stratification exists. However, the converse does not hold; i.e., the presence of an open path allows but does not imply dependency. Independence may arise through cancellation of dependencies. As a consequence, adjacent variables may be marginally independent, such as A and E if the dependencies through paths A → E and A → C → E cancel each other.

3.10.1.2 Statistical implications

A joint probability distribution for the variables in a graph is *compatible* with the graph if two sets of variables are independent given S whenever S separates them. For such distributions, the two sets of variables are statistically unassociated if there is no open path between them. Many special results follow for distributions compatible with a DAG. For example, if X is not an ancestor of any variable in a set T that is part of a DAG, then T and X will be independent given $pa[X]$. Therefore, a distribution compatible with a DAG can be reduced to a product of factors (Markov factorization) $P(x|pa[X])$ with one factor for each variable X in the DAG. As an example, consider a causal Bayesian network whose nodes represent random variables $X = \{x_1, \ldots, x_n\}$.[19] The probability functions for the nodes have values that are dependent on the parent nodes. Hence, if x_i denotes the variable and its corresponding node, $pa[a_i]$ signifies the parents of node x_i, and a_i designates the variables corresponding to those parents, the joint probability distribution for X is given by

$$p(X) = \prod_{i=1}^{n} (p(x_i)|pa[a_i]). \tag{3.76}$$

Suppose that we are interested in the effect of X on Y in a DAG, and we assume a probability model compatible with the DAG. Then, given a sufficient conditioning set S, the only source of association between X and Y within strata of S will be the directed paths from X to Y. Thus, the net effect of $X = x_1$ versus $X = x_0$ on Y when $S = s$ is defined as $P(y|x_1,s) - P(y|x_0,s)$, i.e., the difference in risks of $Y = y$ at $X = x_1$ and $X = x_0$. Other effect measures such as the risk ratio $P(y|x_1,s)$ / $P(y|x_0,s)$ may also be used.

A *standardized effect* is a difference or ratio of weighted averages of these stratum-specific $P(y|x,s)$ over S, using a common weighting distribution. The latter definition can be generalized to include intermediate variables in S by allowing the weighting distribution to causally depend on X. Furthermore, given a set Z of intermediates along all directed paths from X to Y with $X - Z$ and $Z - Y$ unbiased, one can produce formulas for the $X - Y$ effect as a function of the $X - Z$ and $Z - Y$ effects.

When S is sufficient, $P(y|x,s)$ may be identified with the distribution of potential outcomes given S, especially when X is an intervention variable. In this case, sufficiency of a set S implies that the potential outcome distribution equals $\Sigma_s P(y|x,s) P(s)$, which is the risk of $Y = y$ given $X = x$ standardized to the S distribution.

3.10.2 Application to maneuver-based trajectory prediction and criticality assessment

Schreier, Willert, and Adamy[17] applied causal Bayesian networks to develop a maneuver-based trajectory prediction and criticality assessment model that functions as an advanced driver assistance system. The model contains three major parts, namely, maneuver detection, trajectory prediction, and criticality assessment, whose functions are summarized below.

- *Maneuver detection* estimates the current driving maneuver of every vehicle (the ego or reference vehicle and all other vehicles in its vicinity) via Bayesian inference. This is accomplished by modeling a causal

Bayesian network whose inputs are measured vehicle and road configuration data assumed to be available from tracking, mapping, lane recognition, and localization systems. The inference result is a probability mass function (pmf) $p(M_{E,k})$ of the discrete maneuver random variable $M_{E,k}$ for the ego vehicle at each time step k and for the $i \in \{1, \ldots, f\}$ other vehicles $V_{i,k}$ within the observed traffic scene $p(M_{V_{i,k}})$.

- *Trajectory prediction* utilizes maneuver-specific models to project the state $\mathbf{X} = (x, y, \psi)^T$ of each vehicle forward in time within a common global coordinate system, where x is the longitudinal distance traveled along a lane, y is the lateral distance traveled across a lane, and ψ is the yaw angle of the vehicle with respect to the along-lane direction. The prediction process results in a joint probability density function (pdf) $p(x_{k,\ k+1,\ \ldots\ ,\ k+T_P})$ of future states for the $T_P \in N$ prediction time steps for each vehicle.
- *Criticality assessment* combines the individual joint pdfs with a parametric free space (PFS) map-based representation of the static environment having a pdf equal to $p(x_{\mathrm{PFS},k})$ to estimate the collision probability $P[C_k(T_P)]$ of the event that the ego vehicle collides with at least one other vehicle or the static driving environment (e.g., guard rails, sign posts, trees, and light poles) at least once within the *prediction horizon* $\{k, k+1, \ldots, k+T_P\}$ via Monte Carlo simulation. Trajectory samples do not have to be generated for the whole prediction horizon, but only until the first collision is detected because the combined collision event probability does not change in the case of subsequent collisions. Subsequently, the time-to-critical-collision probability (TTCCP) criticality metric is calculated as the required time for the probability of the collision event to exceed a predetermined value. Schreier, Willert, and Adamy[17] use the TTCCP as it generalizes the time-to-collision (TTC) metric in arbitrary uncertain multi-object driving environments and is valid for longer prediction horizons.

3.10.2.1 Maneuver detection

The evidence nodes in the top row of each vehicle model in Fig. 3.10 consist of causal evidence such as left, right, and current lane existence of a vehicle; time

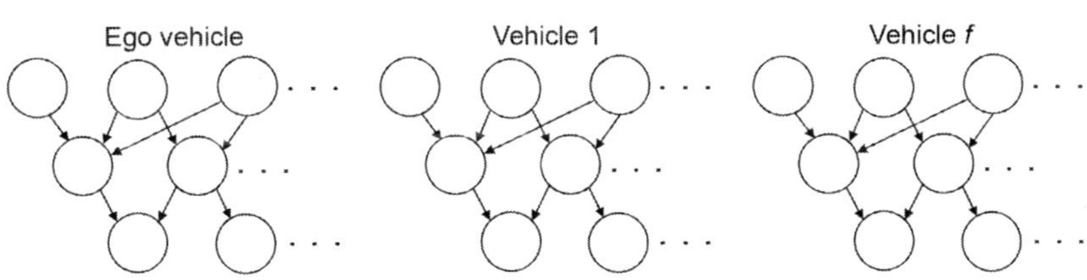

Figure 3.10 Causal Bayesian network for maneuver detection for each vehicle within the traffic scene (notional).

to enter lane on the right or the left; time to turn left or right; left, right turning existence; relative velocity to front object; front object existence; and time to front object. The middle row contains the maneuver layer that has nodes representing each of the maneuver models. The bottom row is the diagnostic layer that models the consequences (symptoms) of the maneuvers in the form of measurable physical motion states, namely, yaw angle with respect to road tangent, lateral acceleration perpendicular to road tangent, lateral velocity perpendicular to road tangent, and longitudinal acceleration along road tangent. At every time step k, the network is updated with the available evidence and the individual maneuver nodes are normalized to form a single, valid, discrete maneuver random variable M with eight maneuver states and corresponding pmfs $p(M)$. The maneuver states are defined by the trajectory prediction models. Note that there are two states each for the lane change and turn models as described in the next section.

3.10.2.2 Trajectory prediction

Each maneuver has a corresponding trajectory prediction model. The prediction models adapt to the traffic scene by incorporating evidence supplied by the vehicle and the environment. The models add random components to account for different driving styles (prediction uncertainties) and are initialized with the current vehicle and environment state pdfs at each time step. The trajectory models are as follows.

1. Follow road (FR): The model predicts the vehicle's longitudinal position along the road according to a (nearly) constant acceleration model described in Section 5.11.
2. Follow vehicle (FV): The model assumes that the driver reacts to another vehicle in front and accelerates comfortably such that an adequate mean reference time gap occurs in the near future. It assumes moderate reaction of the driver, similar to reaction times in adaptive cruise control systems, and does not predict abrupt braking or acceleration maneuvers.
3. Target brake (TB): This model presumes that the vehicle stops in back of a braking target, such as an obstacle, another vehicle, a stop line, or a red traffic signal. The predicted acceleration equals the necessary constant acceleration for the ego vehicle traveling at its current velocity to stop within a specific safety distance from the braking target under some practical minimum acceleration. The safety distance is assumed to be normally distributed about a mean reference safety distance.
4. Lane change (LC): The model predicts lane changes to the left or right according to a prototype model represented by a sine half-cycle in road coordinates that accounts for the longitudinal lane change distance, lane widths, reference maneuver length, and the rate of change of yaw angle.

5. Turn (TU): The model predicts turns to the left or right by partitioning the maneuver into three sections: (1) before the turn on the initial road, (2) the turn itself, and (3) the remaining prediction on the destination road after the turn is complete. The initial and destination roads are assumed straight in the vicinity of the turn, while the destination road is rotated versus the initial road by an angle between 0 and 180 deg.
6. Trash maneuver (TR): This model subsumes all motions not belonging to the specific maneuver classes above, e.g., irrational movements of a drunk driver. The model also allows the seamless transition from highly structured to nonstructured environments. The trash maneuver model predicts the motion according to a constant turn rate and acceleration model having zero-mean, white Gaussian noise. It also incorporates the longitudinal and lateral position of the ego vehicle's yaw angle, yaw rate, and velocity and acceleration in the driving direction.

3.10.2.3 Maneuver model combination and criticality assessment

After transformation to a common, global reference frame, every maneuver prediction model generates a joint pdf $p(x_{k,\ k+1,\ \ldots,\ k+T_p}|M_{j,k})$ over the prediction horizon, where the state of the ego vehicle and other vehicles is described by a vector of the form $\mathbf{X} = (x,\ y,\ \psi)^{\mathrm{T}}$ for all $j \in \{1,\ \ldots,\ r\}$ maneuvers of each vehicle.

By marginalizing out the maneuver states from the combined pdf $p(x_{E,\ k,\ k+1,\ \ldots,\ k+T_p},\ M_k)$ with the expressions for $p(M_{E,k})$ and $p(M_{V_{i,k}})$ that were obtained from the maneuver detection process, we find that

$$p(x_{E,k,k+1,\ldots,k+T_P}) = \sum_{j=1}^{r} p(x_{E,k,k+1,\ldots,k+T_P}|M_{E,k} = j)p(M_{E,k} = j), \quad (3.77)$$

and

$$p(x_{V_i,\ k,\ k+1,\ \ldots,\ k+T_P}) = \sum_{j=1}^{r} p(x_{V_i,k,\ k+1,\ \ldots,\ k+T_P}|M_{V_i,k} = j)p(M_{V_i,k} = j) \quad (3.78)$$

for the ego vehicle and the other $i \in \{1,\ \ldots\ ,f\}$ vehicles, respectively. These densities, used for criticality assessment, are thus a combination of all prediction models that are weighted with their current maneuver probabilities. Their product replaces the joint density distribution for the ego vehicle and other vehicles since the trajectories of all of the vehicles are assumed to be independent of each other.

The last step before finding the TTCCP is to combine the pdfs in Eqs. (3.77) and (3.78) with the PFS map that represents the static objects in the driving environment. Once completed, this action leads to the calculation of the probability of the combined event $C_k(T_P)$ that the ego vehicle collides with at least one vehicle or an obstacle in the static driving environment at least once within the prediction horizon as a function of T_P (the length of the prediction horizon, or equivalently, the number of prediction time steps).

The criticality metric TTCCP is subsequently defined as the time until the combined collision probability $P[C_k(T_P)]$ exceeds a critical collision probability (CCP). Therefore,

$$\text{TTCCP} = \min(T_P | P[C_k(T_P)] > \text{CCP}) \times T, \tag{3.79}$$

where T is the prediction sampling time.

The TTCCP monotonically increases over the prediction time and gives the collision probability over a future time span and not the collision probability at future time instants. The combined collision probability $P[C_k(T_P)]$ is found through Monte Carlo simulation.[17]

3.11 Summary

Bayesian inference determines the likelihoods that unknown states H_i are present, where the new likelihoods associated with the H_i are computed from the product of the original prior information and the information gained by observation (the evidence). This is encoded in the posterior probability $P(H_i|E)$ that describes the likelihoods associated with H_i given the evidence E. In recursive operation of the Bayesian process, the original priors are replaced by the posterior probability computed at the previous time step. Updating of the posterior belief does not have to be delayed until evidence from all sources is collected, but can be implemented incrementally as a sensor or other information source reports.

In a data fusion application of Bayesian inference, each sensor provides an identity declaration D or hypothesis about the object's identity based on observations and a sensor-specific classification algorithm. The algorithm provides estimates of the probability that the sensor will declare the object to be a certain type, given the object is in fact type i. This probability is denoted by $P(D|O_i)$. Sensor declarations are combined using a generalization of Bayes' theorem to produce an updated, joint probability for each entity O_i.

The method requires:

- The ability to compute or model the likelihood functions $P(E|H_i)$ for each sensor and object in the scenario of interest,
- *a priori* probabilities that hypotheses $P(H_i)$ are true,
- That when *a priori* information about $P(H_i)$ is lacking, the principle of indifference is invoked; i.e., initial values for $P(H_i)$ for all i are set equal to each other,
- Mutual exclusivity of competing hypotheses,

- Discarding of past evidence or data once the posterior probability $P(H_i|E)$ is calculated, and
- Conditionally independent sensors.

Several applications of Bayesian inference were discussed to show the practical implementation of this method to ITS. These were identification of the location of an incident or a vehicle using multiple-sensor data and the likelihood functions of the sensors given a hypothesis; incident detection using an influence diagram and joint sensor report with data from roadway sensors, cellular telephone calls, and commercial truck radio reports; truck classification according to a 14-class FHWA scheme using a naïve Bayes classifier to combine the individual classifications from several trained classifiers; and the application of causal Bayesian networks to vehicle maneuver detection and the prediction of the time span over which a collision will occur due to vehicle maneuvers.

Caution: Bayesian inference does not have a convenient representation for ignorance or uncertainty. Therefore, it equates concepts of ignorance and belief when direct knowledge is lacking. This limitation is overcome with Dempster–Shafer evidential reasoning, which is described in the next chapter.

References

1. H. F. Durrant-Whyte, *Introduction to Estimation and the Kalman Filter*, V2.2, Jan. 2001. https://www.isip40.it/resources/Dispense/DataFusionSIIT/Estimation_KF.pdf. Accessed Aug. 6, 2018.
2. H. F. Durrant-Whyte, *Multi Sensor Data Fusion*, V1.2, Jan. 2001. https://www.mit.bme.hu/system/files/oktatas/targyak/7357/MultiSensorDataFusionNotes_DurrantWhyte_2011.pdf. Accessed Dec. 3, 2018.
3. H. Durrant-Whyte and T. Henderson, "Multisensor data fusion," *Handbook of Robotics, Part C—Sensing and Perception*, O. Khatib and B. Siciliano, Eds., Springer, New York, Chapter 25 (2008).
4. L. A. Klein, *Sensor and Data Fusion: A Tool for Information Assessment and Decision Making*, Second Edition, SPIE Press, Bellingham, Washington, (2012) [doi: 10.1117/3.928035].
5. P. G. Hoel, *Introduction to Mathematical Statistics*, John Wiley & Sons, New York (1947).
6. H. Cramér, *Mathematical Methods of Statistics*, Ninth Printing, Princeton University Press, Princeton (1961).
7. L. L. Chao, *Statistics: Methods and Analyses*, McGraw-Hill Book Company, New York (1969).
8. J. Pearl, *Probabilistic Reasoning in Intelligent Systems: Networks of Plausible Inference*, Morgan Kaufmann Publishers, San Mateo, California (1988).

9. E. Waltz and J. Llinas, *Multisensor Data Fusion*, Artech House, Norwood, Massachusetts (1990).
10. L. A. Klein, *Sensor Technologies and Data Requirements for ITS*, Artech House, Norwood, Massachusetts (2001).
11. S. V. Hernandez, A. Tok, and S. G. Ritchie, "Integration of weigh-in-motion (WIM) and inductive signature data for truck body classification," *Transportation Research Part C* **68**, 1–21 (2016).
12. P. Domingos and M. Pazzani, "On the optimality of the simple Bayesian classifier under zero-one loss," *Machine Learning* **29**(2–3), 103–130 (1997).
13. N. Van Dan and M. Kameyama, "Bayesian-networks-based motion estimation for a highly safe intelligent vehicle," *SICE-ICASE International Joint Conference*, Bexco, Busan, S. Korea, Oct. 18–21 (2006).
14. G. Pavlin, M. Maris, and F. Groen, "Causal Bayesian networks for robust and efficient fusion of information obtained from sensors and humans," *2007 IEEE Instrumentation and Measurement Technology Conference (IMTC 2007)*, Warsaw, May 1–3 (2007).
15. R. Maarefdoust and S. Rahati, "Traffic modeling with multi-agent Bayesian and causal networks and performance prediction for changed setting system," *IEEE 2010 Second International Conference on Machine Learning and Computing*, Bangalore, Feb. 9–11, pp. 17–21 (2010).
16. Y. Zhou and Y. Zhang, "Applications of Bayesian network in fault diagnosis of braking deviation system," *IEEE Fourth International Symposium on Computational Intelligence and Design*, Hangzhou, Oct. 28–30 (2011).
17. M. Schreier, V. Willert, and J. Adamy, "An integrated approach to maneuver-based trajectory prediction and criticality assessment in arbitrary road environments," *IEEE Trans. on Intelligent Transportation Systems* **17**(10), 2751–2766 (2016).
18. S. Greenland and J. Pearl, "Causal Diagrams," Technical Report R-332, prepared for *Encyclopedia of Epidemiology*, Jun. (2006).
19. X. Liu and S. Tan, "Identifying causal relationships in an urban information modeling framework," *2010 International Joint Conference on Neural Networks (IJCNN)*, Barcelona, July 18–23 (2010).

Chapter 4
Dempster–Shafer Evidential Reasoning for Traffic Management

Dempster–Shafer evidential reasoning, a probability-based data fusion classification algorithm, is useful when the sensors (or, more generally, the information sources) contributing data cannot associate a 100% probability of certainty to their output decisions. The algorithm captures and combines whatever certainty exists in the object discrimination capability of the sensors. The sensors must act as independent sources when offering evidence for a proposition, i.e., the presence of one or more objects or events. The information provided by two or more sensors is combined using Dempster's rule to find the intersection or conjunction of the individual propositions reported by the sensors and their associated probabilities, which are expressed as probability masses. When the intersection of the propositions reported by the sensors is a null or empty set, Dempster's rule redistributes the conflicting probability to the nonempty set elements. In the event that the conflicting probability is large, application of Dempster's rule can lead to counterintuitive conclusions. Several modifications to the original Dempster–Shafer theory have been proposed to accommodate these situations. Perhaps the most difficult part of applying Dempster–Shafer theory in its original or modified forms is obtaining probability mass functions. Five methods for creating these functions are discussed. An example is given of the application of Dempster–Shafer reasoning to create a decision support system that enhances a traffic manager's understanding of the conditions giving rise to the collected data.

4.1 Overview of the Process

Figure 4.1 describes the Dempster–Shafer fusion process for identifying objects from multisensor data.[1] Each sensor has a set of observables corresponding to the phenomena that generate data received about objects and their

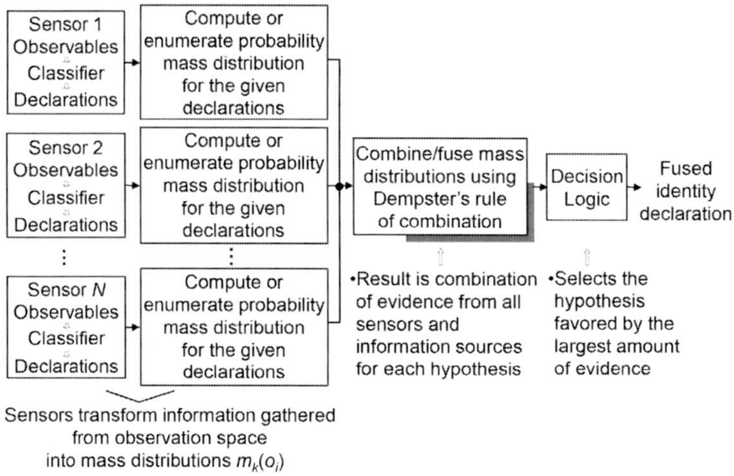

Figure 4.1 Dempster–Shafer fusion process (adapted from Ref. 1).

surroundings. In this diagram, a sensor operates on the observables with its particular set of object classification algorithms. The knowledge gathered by each sensor k, where $k = 1, \ldots, N$, associates a declaration of object type (referred to in the figure by object o_i, where $i = 1, \ldots, n$) with a probability mass or basic probability assignment $m_k(o_i)$ between 0 and 1. The probability mass expresses the certainty of the declaration or hypothesis, i.e., the amount of support or belief attributed *directly* to the declaration. Probability masses closer to unity characterize decisions made with more definite knowledge or less uncertainty about the nature of the object.

The probability masses for the decisions made by each sensor are then combined using Dempster's rule of combination. The hypothesis favored by the largest accumulation of evidence from all contributing sensors is selected as the most probable outcome of the fusion process. A computer stores the relevant information from each sensor concerning an object's identity and probability mass. The converse is also true, namely, that objects not supported by evidence from any sensor are not stored.

In addition to real-time sensor data, other information or rules can be stored in the information base to improve the overall decision or object discrimination capability. Examples of such rules are "Ships detected in known shipping lanes are cargo vessels" and "Objects in previously charted Earth orbits are weather or reconnaissance satellites."

4.2 Implementation of the Method

Assume that a set of n mutually exclusive and exhaustive propositions exists; e.g., an object is of type $a_1, a_2, \ldots,$ or a_n. This is the set of all propositions making up the hypothesis space, called the frame of discernment, and is denoted by θ. Probability mass $m(a_i)$ assignments may be made to any of the original propositions, to the union of the propositions, and to the negation of a proposition based on available sensor information. As an example of a union or disjunction, consider that the object type a_1 or a_2 (denoted by $a_1 \cup a_2$) is assigned probability mass $m(a_1 \cup a_2)$ by a sensor.

A proposition is called a focal element if its probability mass is greater than zero. The number of combinations of propositions that exists (including all possible unions and θ itself, but excluding the null set) equals $2^n - 1$. For example, if $n = 3$, there are $2^3 - 1 = 7$ propositions given by a_1, a_2, a_3, $a_1 \cup a_2$, $a_1 \cup a_3$, $a_2 \cup a_3$, and $a_1 \cup a_2 \cup a_3$, where θ equals $a_1 \cup a_2 \cup a_3$. When the frame of discernment contains n focal elements, the power set consists of 2^n elements, including the null set.

In the event that all of the probability mass cannot be *directly* assigned by the sensor to any of the propositions, their unions, or the negation of a proposition such as $m(\bar{a}_1) = m(a_2 \cup a_3 \cup \ldots \cup a_n)$ [a raised bar denotes the negation of a proposition], the remaining mass is assigned to the frame of discernment θ (representing uncertainty as to further definitive assignment) as $m(\theta) = m(a_1 \cup a_2 \cup \ldots \cup a_n)$. The mass assigned to θ represents the uncertainty that the sensor has concerning the accuracy and interpretation of the evidence.[2] The sum of probability masses over all propositions, uncertainty, and negation equals unity.

To illustrate these concepts, suppose that two sensors observe a scene in which there are three objects. Sensor A identifies the object as belonging to one of the three possible types: a_1, a_2, or a_3. Sensor B declares the object to be of type a_1 with a certainty of 80%. The intersection of the data from the two sensors is written as

$$(a_1 \text{ or } a_2 \text{ or } a_3) \text{ and } (a_1) = (a_1), \tag{4.1}$$

or upon rewriting with mathematical symbols replacing the words 'or' and 'and' as

$$(a_1 \cup a_2 \cup a_3) \cap (a_1) = (a_1). \tag{4.2}$$

Only a probability of 0.8 can be assigned to the intersection of the sensor data based on the 80% confidence associated with the output from Sensor B. The remaining probability of 0.2 is assigned to uncertainty represented by the union (disjunction) of (a_1 or a_2 or a_3).[3]

4.3 Support, Plausibility, and Uncertainty Interval

According to Shafer, "an adequate summary of the impact of the evidence on a particular proposition a_i must include at least two items of information: a report on how well a_i is supported and a report on how well its negation $\bar{a}_i$ is supported."[4] These two items of information are conveyed by the proposition's degree of support and its degree of plausibility.

Support (also referred to as belief) for a given proposition is defined as the sum of all masses assigned *directly* by the sensor to that proposition or its subsets.[4,5] A subset is called a focal subset if it contains elements of θ with mass greater than zero. Thus, the support for object type a_1, denoted by $S(a_1)$, contributed by a sensor is

$$S(a_1) = m(a_1). \tag{4.3}$$

Support for the proposition that the object is either type a_1, a_2, or a_3 is

$$\begin{aligned} S(a_1 \cup a_2 \cup a_3) &= m(a_1) + m(a_2) + m(a_3) + m(a_1 \cup a_2) + m(a_1 \cup a_3) \\ &\quad + m(a_2 \cup a_3) + m(a_1 \cup a_2 \cup a_3). \end{aligned} \tag{4.4}$$

Plausibility of a given proposition is defined as the sum of all mass not assigned to its negation.[4] Consequently, plausibility defines the mass that is free to move to the support of a proposition. The plausibility of a_i, denoted by $Pl(a_i)$, is written as

$$Pl(a_i) = 1 - S(\bar{a}_i), \tag{4.5}$$

where $S(\bar{a}_i)$ is called the dubiety and represents the degree to which the evidence impugns a proposition, i.e., supports the negation of the proposition.

Plausibility can also be computed as the sum of all masses belonging to subsets a_j that have a non-null intersection with a_i. Accordingly,

$$Pl(a_i) = \sum_{a_j \cap a_i \neq 0} m(a_j). \tag{4.6}$$

Thus, when $\theta = \{a_1, a_2, a_3\}$, the plausibility of a_1 is computed as the sum of all masses compatible with a_1, which includes all unions containing a_1 and θ such that

$$Pl(a_1) = m(a_1) + m(a_1 \cup a_2) + m(a_1 \cup a_3) + m(a_1 \cup a_2 \cup a_3). \tag{4.7}$$

The more general equation for calculating plausibility is Eq. (4.5), as Eq. (4.6) is only valid when values for the probability masses of all the subsets are known.

An uncertainty interval is defined by $[S(a_i), Pl(a_i)]$, where

$$S(a_i) \leq Pl(a_i). \tag{4.8}$$

The Dempster–Shafer uncertainty interval displayed in Figure 4.2 illustrates the concepts just discussed.[6,7] The lower bound or support for a proposition

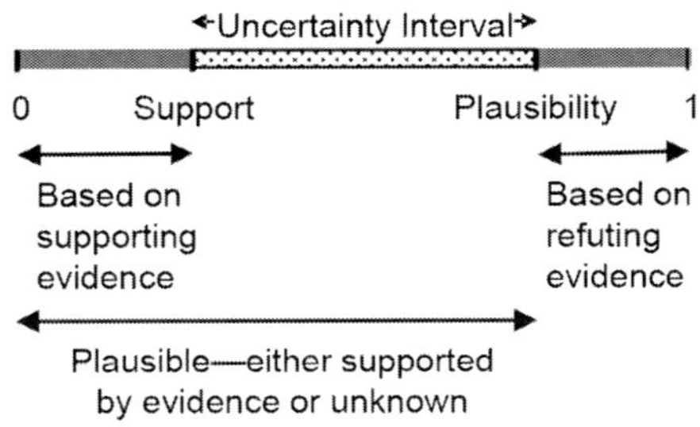

Figure 4.2 Dempster–Shafer uncertainty interval for a proposition.

equals the minimal commitment for the proposition based on *direct* sensor evidence. The upper bound or plausibility equals the support plus any potential commitment. Therefore, these bounds show what proportion of evidence is truly in support of a proposition and what proportion results merely from ignorance; alternatively, they show the requirement to normalize the sum of the probability masses to unity. As the examples that follow demonstrate, the support for a hypothesis a and the support for its negation $\bar{a}$ do not necessarily sum to 1, which is a major difference with probability theory.[8]

Support and probability mass obtained from a sensor (knowledge source) represent different concepts. Support is calculated as the sum of the probability masses that *directly* confirm or support the proposition and its unions. Probability mass is determined from the sensor's ability to assign some certainty to a proposition based on the evidence.

Table 4.1 provides further interpretations of uncertainty intervals. For example, the uncertainty interval [0, 1] represents total ignorance about proposition a_i since there is no direct support for a_i, but also no refuting evidence, as can be seen by calculating the plausibility $1 - S(\bar{a}_i) = 1$ since $S(\bar{a}_i) = 0$. Here the plausible range equals unity, as does the uncertainty interval. The uncertainty interval denoted by [0.6, 0.6] contains equal support and plausibility values. It indicates a definite probability of 0.6 for proposition a_i since both the direct support and plausibility are 0.6. In this case, the uncertainty interval equals zero. Support and plausibility values represented by

Table 4.1 Interpretation of uncertainty intervals for proposition a_i.

Uncertainty Interval $[S(a_i), Pl(a_i)]$	Interpretation
[0, 1]	Total ignorance about proposition a_i.
[0.6, 0.6]	A definite probability of 0.6 for proposition a_i.
[0, 0]	Proposition a_i is false.
[1, 1]	Proposition a_i is true.
[0.25, 1]	Evidence provides partial support for proposition a_i.
[0, 0.85]	Evidence provides partial support for $\bar{a}_i$.
[0.25, 0.85]	Probability of a_i is between 0.25 and 0.85; i.e., the evidence simultaneously provides support for both a_i and $\bar{a}_i$.

[0, 0] indicate that the proposition a_i is false, as a plausibility of zero implies that all of the probability mass is assigned to the negation of a_i. This becomes apparent when writing the plausibility as $1 - S(\bar{a}_i)$, which equals zero since $S(\bar{a}_i) = 1$. When a_i is known to be true, [1, 1] represents the support and plausibility values. The uncertainty interval is zero since all of the probability mass is assigned to the proposition a_i. Therefore, the support for a_i is 1, and the plausibility, $1 - S(\bar{a}_i)$, is also 1 since the plausibility of 1 implies that $S(\bar{a}_i) = 0$.

The support and plausibility values [0.25, 1] imply evidence that partially supports proposition a_i with a support value of 0.25. A plausibility of one indicates that there is not any direct evidence to refute a_i. All of the probability mass in the uncertainty interval of length 0.75 is free to move to the support of a_i. The interval [0, 0.85] implies partial support for the negation of a_i since there is no direct evidence to support a_i while there is partial evidence to support $\bar{a}_i$; i.e., $S(\bar{a}_i) = 0.15$. The support and plausibility represented by [0.25, 0.85] show partial direct support for a_i and partial direct support for its negation as $S(\bar{a}_i) = 0.15$. In this case, the uncertainty interval represents probability mass that is available to move to support a_i or $\bar{a}_i$.

As an example of how the uncertainty interval is computed from the information a sensor provides, consider once again a single sensor denoted as Sensor A that gathers data concerning three propositions a_1, a_2, and a_3. The frame of discernment θ is given by

$$\theta = \{a_1, a_2, a_3\}. \tag{4.9}$$

The negation of proposition a_1 is represented by

$$\bar{a}_1 = \{a_2, a_3\}. \tag{4.10}$$

Assume that probability masses are contributed by Sensor A to the propositions a_1, $\bar{a}_1$, $a_1 \cup a_2$, and θ as

$$m_A(a_1, \bar{a}_1, a_1 \cup a_2, \theta) = (0.4, 0.2, 0.3, 0.1). \tag{4.11}$$

Table 4.2 shows the uncertainty intervals for a_1, $\bar{a}_1$, $a_1 \cup a_2$, and θ calculated using these numerical values. The uncertainty interval computations for a_1 and

Table 4.2 Uncertainty interval calculation for propositions a_1, $\bar{a}_1$, $a_1 \cup a_2$, and θ.

Proposition	Support $S(a_i)$	Plausibility $1 - S(\bar{a}_1)$	Uncertainty Interval
a_1	0.4 (given)	$1 - S(\bar{a}_1) = 1 - 0.2 = 0.8$	[0.4, 0.8]
$\bar{a}_1$	0.2 (given)	$1 - S(a_1) = 1 - 0.4 = 0.6$	[0.2, 0.6]
$a_1 \cup a_2$	$S(a_1) + S(a_1 \cup a_2)$ $= 0.4 + 0.3 = 0.7$	$1 - S(\bar{a}_1 \cup \bar{a}_2)$ $= 1 - S(\bar{a}_1 \cap \bar{a}_2) = 1 - 0 = 1$*	[0.7, 1]
θ	$S(\theta) = 1$	$1 - S(\bar{\theta}) = 1 - 0 = 1$	[1, 1]

*Only probability mass assigned directly by Sensor A to $\bar{a}_1 \cap \bar{a}_2$ is used in the calculation. Because Sensor A has not assigned any probability mass directly to $\bar{a}_1 \cap \bar{a}_2$, the support for $\bar{a}_1 \cap \bar{a}_2$ is zero. Thus, the plausibility of $a_1 \cup a_2$ is unity.

$\bar{a}_1$ are straightforward since they are based on direct sensor assignment of probability mass. The uncertainty interval for proposition $a_1 \cup a_2$ is found using the direct evidence from Sensor A that supports a_1 and $a_1 \cup a_2$. The probability mass $m_A(\theta)$, i.e., the mass not assignable to a smaller set of propositions, is not included in any of the supporting or refuting evidence for $a_1 \cup a_2$ because $m_A(\theta)$ represents the residual uncertainty of the sensor in distributing the remaining probability mass directly to any other propositions or unions in θ based on the evidence. That is, the evidence has allowed the sensor to assign *direct* probability mass only to propositions a_1, $\bar{a}_1$, and $a_1 \cup a_2$. The remaining probability mass assigned to $m_A(\theta)$ implies that it is distributed in some unknown manner among the totality of all propositions. Support for θ equals unity because θ is the totality of all propositions; plausibility for θ also equals unity because support is not assigned outside of θ. Therefore, $m_A(\bar{\theta}) = 0$, and $Pl(\theta) = 1 - S(\bar{\theta}) = 1 - 0 = 1$.

As a final example of uncertainty interval calculation, consider that Sensor A now provides information about the identification of propositions a_1, a_2, a_3 such that $\theta = \{a_1, a_2, a_3\}$ and $\bar{a}_2$ is represented by the set $\{a_1, a_3\}$. The probability mass assignments are $m_A(a_1, a_2 \cup a_3, a_3, \bar{a}_2) = (0.4, 0.2, 0.1, 0.3)$. Table 4.3 presents the results of the uncertainty interval calculation for all propositions assigned probability mass by Sensor A and for a_2.

4.4 Dempster's Rule for Combining Multiple-Sensor Data

Dempster's rule supplies the formalism to combine the probability masses provided by multiple sensors or information sources for compatible propositions. The output of the fusion process is given by the intersection of the propositions having the largest probability mass. Propositions are compatible when their intersection exists. Dempster's rule also treats intersections that form a null set, i.e., incompatible propositions. In this case, the rule adjusts the probability masses associated with null intersections to zero and increases the probability masses of the nonempty set intersections by a normalization factor K such that their sum is unity.

Table 4.3 Uncertainty interval calculation for propositions a_1, $a_2 \cup a_3$, a_3, $\bar{a}_2$, and a_2.

Proposition	Support $S(a_i)$	Plausibility $1 - S(\bar{a}_i)$	Uncertainty Interval
a_1	0.4 (given)	$1 - S(\bar{a}_1) = 1 - 0 = 1$	[0.4, 1]
$a_2 \cup a_3$	$m_A(a_2) + m_A(a_3) + m_A(a_2 \cup a_3)$ $= 0 + 0.1 + 0.2 = 0.3$	$1 - S(\bar{a}_2 \cup \bar{a}_3)$ $= 1 - S(\bar{a}_2 \cap \bar{a}_3)$ $= 1 - 0 = 1$	[0.3, 1]
a_3	$S(a_3) = 0.1$ (given)	$1 - S(\bar{a}_3) = 1 - 0 = 1$	[0.1, 1]
$\bar{a}_2$	$S(\bar{a}_2) = 0.3$ (given)	$1 - S(a_2) = 1 - 0 = 1$	[0.3, 1]
a_2	$S(a_2) = 0$	$1 - S(\bar{a}_2) = 1 - 0.3 = 0.7$	[0, 0.7]

The general form of Dempster's rule for the total probability mass committed to an event c defined by the combination of evidence $m_A(a_i)$ and $m_B(b_j)$ from Sensors A and B is given by

$$m(c) = K \sum_{a_i \cap b_j = c} [m_A(a_i) m_B(b_j)], \tag{4.12}$$

where $m_A(a_i)$ and $m_B(b_j)$ are probability mass assignments on θ,

$$K^{-1} = 1 - \sum_{a_i \cap b_j = \phi} [m_A(a_i) m_B(b_j)], \tag{4.13}$$

and ϕ is defined as the null set. If K^{-1} is zero, then m_A and m_B are completely contradictory, and the sum defined by Dempster's rule does not exist. The probability mass calculated in Eq. (4.12) is termed the orthogonal sum and is denoted by $m_A(a_i) \oplus m_B(b_j)$.

4.5 Vehicle Detection Using Dempster–Shafer Evidential Reasoning

Applications of the Dempster–Shafer technique and Dempster's rule to vehicle detection are illustrated with a four-vehicle, two-sensor example. The first problem applies Dempster's rule to data that does not produce a null set. The second example contains sensor data that leads to a null set. The third case includes sensor data that report only singleton events and can, therefore, be solved with Dempster–Shafer or Bayesian methods.

4.5.1 Dempster's rule applied to compatible data sets

Suppose that two vehicle types and locations are possible, namely,

$$\begin{bmatrix} a_1 = \text{commercial truck in lane 1} & a_3 = \text{commercial truck in lane 2} \\ a_2 = \text{passenger car in lane 1} & a_4 = \text{passenger car in lane 2} \end{bmatrix}.$$

Furthermore, Sensor A provides truck classification data according to the probability mass matrix given by

$$m_A = \begin{bmatrix} m_A(a_1 \cup a_3) = 0.6 \\ m_A(\theta) = 0.4 \end{bmatrix}, \tag{4.14}$$

where $m_A(\theta)$ represents the uncertainty associated with the rules that determine that the vehicle is a truck.

Sensor B is able to classify lane 2 vehicles with the probabilities

$$m_B = \begin{bmatrix} m_B(a_3 \cup a_4) = 0.7 \\ m_B(\theta) = 0.3 \end{bmatrix}, \tag{4.15}$$

where $m_B(\theta)$ represents the uncertainty associated with the rules that determine that the vehicle is in lane 2.

Dempster's rule is implemented by forming a matrix with the probability masses that are to be combined entered along the first column and last row as illustrated in Table 4.4. Inner matrix (row, column) elements are computed as the product of the probability mass in the same row of the first column and the same column of the last row. The proposition corresponding to an inner matrix element equals the intersection of the propositions that are multiplied. Accordingly, matrix element (1, 2) represents the proposition formed by the intersection of uncertainty θ from Sensor A and $(a_3 \cup a_4)$ from Sensor B, namely, that a vehicle (truck or car) is in lane 2. The probability mass $m(a_3 \cup a_4)$ associated with the intersection of these propositions is

$$m(a_3 \cup a_4) = m_A(\theta)m_B(a_3 \cup a_4) = (0.4)\,(0.7) = 0.28. \tag{4.16}$$

Matrix element (1, 3) represents the intersection of the uncertainty propositions from Sensor A and Sensor B. The probability mass $m(\theta)$ associated with the uncertainty intersection is

$$m(\theta) = m_A(\theta)m_B(\theta) = (0.4)\,(0.3) = 0.12. \tag{4.17}$$

Matrix element (2, 2) represents the proposition formed by the intersection of $(a_1 \cup a_3)$ from Sensor A and $(a_3 \cup a_4)$ from Sensor B, namely, that the vehicle is a commercial truck in lane 2. The probability mass $m(a_3)$ associated with the intersection of these propositions is

$$m(a_3) = m_A(a_1 \cup a_3)m_B(a_3 \cup a_4) = (0.6)\,(0.7) = 0.42. \tag{4.18}$$

Matrix element (2, 3) represents the proposition formed by the intersection of $(a_1 \cup a_3)$ from Sensor A and θ from Sensor B. Accordingly, the probability mass associated with this element is

$$m(a_1 \cup a_3) = m_A(a_1 \cup a_3)m_B(\theta) = (0.6)\,(0.3) = 0.18 \tag{4.19}$$

Table 4.4 Application of Dempster's rule.

First column			
$m_A(\theta) = 0.4$	$m(a_3 \cup a_4) = 0.28$	$m(\theta) = 0.12$	
$m_A(a_1 \cup a_3) = 0.6$	$m(a_3) = 0.42$	$m(a_1 \cup a_3) = 0.18$	
	$m_B(a_3 \cup a_4) = 0.7$	$m_B(\theta) = 0.3$	*Last row*

and corresponds to the proposition that the vehicle is a commercial truck in either lane 1 or lane 2.

The proposition represented by $m(a_3)$, namely, a commercial truck in lane 2, has the highest probability mass in the matrix. Thus, it is typically the proposition selected as the output to represent the fusion of the evidence from Sensors A and B. Note that the inner matrix element values sum to unity.

When three or more sensors contribute information, the application of Dempster's rule is repeated using the inner elements calculated from the first application of the rule as the new first column and the probability masses from the next sensor as the entries for the last row (or *vice versa*). This follows from the commutation and association properties of the orthogonal sum.[4]

4.5.2 Dempster's rule with null set elements

When the intersection of the propositions that define the inner matrix elements form a null set, the probability mass of the null set elements is fixed at zero and the probability mass assigned to the nonempty set elements is increased by the normalization factor K. To illustrate this process, suppose that Sensor B had identified objects 2 and 4 instead of objects 3 and 4, with probability mass assignments given by $m_B{}'$ as

$$m_B{}' = \begin{bmatrix} m_B{}'(a_2 \cup a_4) = 0.5 \\ m_B{}'(\theta) = 0.5 \end{bmatrix}. \tag{4.20}$$

Application of Dempster's rule gives the results in Table 4.5, where element (2, 2) now belongs to the null set. Since probability mass is initially attributed to ϕ, we find the value K that redistributes this mass to the nonempty set members by calculating

$$K^{-1} = 1 - 0.30 = 0.70, \tag{4.21}$$

and its inverse K as

$$K = 1.429. \tag{4.22}$$

Table 4.6 shows the probability mass corresponding to the null set element set equal to zero and the probability masses of the nonempty set elements multiplied by K such that their sum is unity. In this example, a commercial truck is declared present, but its location by lane is undetermined.

Table 4.5 Application of Dempster's rule with a null set.

$m_A(\theta) = 0.4$	$m(a_2 \cup a_4) = 0.20$	$m(\theta) = 0.20$
$m_A(a_1 \cup a_3) = 0.6$	$m(\phi) = 0.30$	$m(a_1 \cup a_3) = 0.30$
	$m_B{}'(a_2 \cup a_4) = 0.5$	$m_B{}'(\theta) = 0.5$

Table 4.6 Probability masses of nonempty set elements increased by K.

$m_A(\theta) = 0.4$	$m(a_2 \cup a_4) = 0.286$	$m(\theta) = 0.286$
$m_A(a_1 \cup a_3) = 0.6$	0	$m(a_1 \cup a_3) = 0.429$
	$m_B{}'(a_2 \cup a_4) = 0.5$	$m_B{}'(\theta) = 0.5$

4.5.3 Dempster's rule with singleton propositions

When probability mass assignments are provided by sensors that report unique singleton events (that is, probability mass is not assigned to the union of propositions or the uncertainty class), the problem becomes Bayesian and the number of null set elements increases as shown in the following example. Assume that two possible vehicle types and locations are present as before, namely,

$$\begin{bmatrix} a_1 = \text{commercial truck in lane 1} & a_3 = \text{commercial truck in lane 2} \\ a_2 = \text{passenger car in lane 1} & a_4 = \text{passenger car in lane 2} \end{bmatrix}.$$

Only now Sensor A's probability mass matrix is given by

$$m_A = \begin{bmatrix} m_A(a_1) = 0.35 \\ m_A(a_2) = 0.06 \\ m_A(a_3) = 0.35 \\ m_A(a_4) = 0.24 \end{bmatrix}, \tag{4.23}$$

and Sensor B's probability mass matrix is given by

$$m_B = \begin{bmatrix} m_B(a_1) = 0.10 \\ m_B(a_2) = 0.44 \\ m_B(a_3) = 0.40 \\ m_B(a_4) = 0.06 \end{bmatrix}. \tag{4.24}$$

Note that the probability mass assignments by each of the sensors sum to unity. Next apply Dempster's rule to compute the orthogonal sum of the m_A and m_B values as indicated in Table 4.7. The only commensurate elements are

Table 4.7 Application of Dempster's rule with singleton events.

$m_A(a_1) = 0.35$	$m(a_1) = 0.035$	$m(\phi) = 0.154$	$m(\phi) = 0.140$	$m(\phi) = 0.021$
$m_A(a_2) = 0.06$	$m(\phi) = 0.006$	$m(a_2) = 0.0264$	$m(\phi) = 0.024$	$m(\phi) = 0.0036$
$m_A(a_3) = 0.35$	$m(\phi) = 0.035$	$m(\phi) = 0.154$	$m(a_3) = 0.140$	$m(\phi) = 0.021$
$m_A(a_4) = 0.24$	$m(\phi) = 0.024$	$m(\phi) = 0.1056$	$m(\phi) = 0.096$	$m(a_4) = 0.0144$
	$m_B(a_1) = 0.10$	$m_B(a_2) = 0.44$	$m_B(a_3) = 0.40$	$m_B(a_4) = 0.06$

those along the diagonal. All others are null set members. The value of K that redistributes the probability mass of null set members to nonempty set propositions is found from

$$\begin{aligned} K^{-1} &= 1 - 0.006 - 0.035 - 0.024 - 0.154 - 0.154 \\ &\quad - 0.1056 - 0.140 - 0.024 - 0.096 - 0.021 \\ &\quad - 0.0036 - 0.021 = 1 - 0.7842 = 0.2158 \end{aligned} \tag{4.25}$$

as

$$K = 4.6339. \tag{4.26}$$

Table 4.8 contains the resulting probability mass matrix having diagonal elements whose orthogonal sum now equals unity. In this case, the most likely event is the presence of a commercial truck in lane 2.

4.6 Singleton Proposition Vehicle Detection Problem Solved with Bayesian Inference

Since probability mass assignments to the uncertainty interval and the union of propositions are zero, we can reformulate the last four-object example in Bayesian terms. The likelihood vector $\mathbf{\Lambda}$ is computed as the scalar product of the likelihood functions $\boldsymbol{\lambda}^1$ and $\boldsymbol{\lambda}^2$ (in this example, these are the probability mass assignments from each sensor). Accordingly,

$$\lambda^1 = P(E^{\mathrm{A}}|H_i) = (0.35,\ 0.06,\ 0.35,\ 0.24), \tag{4.27}$$

$$\lambda^2 = P(E^{\mathrm{B}}|H_i) = (0.10,\ 0.44,\ 0.40,\ 0.06), \tag{4.28}$$

and

$$\Lambda = \lambda^1 \cdot \lambda^2 = (0.035,\ 0.0264,\ 0.140,\ 0.0144). \tag{4.29}$$

From Eq. (3.48),

$$\begin{aligned} P(H_i|E^{\mathrm{A}},\ E^{\mathrm{B}}) &= \alpha\,(0.035,\ 0.0264,\ 0.140,\ 0.0144) \\ &= (0.1622,\ 0.1223,\ 0.6487,\ 0.0667), \end{aligned} \tag{4.30}$$

where $\alpha = 1/(0.035 + 0.0264 + 0.140 + 0.0144) = 4.6339$, which is the same value calculated for K with the Dempster–Shafer method. The result in Eq. (4.30) is identical to that obtained with the Dempster–Shafer method in Table 4.8, namely, that the most likely event is a_3 (commercial truck in lane 2).

Table 4.8 Redistribution of probability mass to nonempty set elements through multiplication by *K*.

$m_A(a_1) = 0.35$	$m(a_1) = 0.1622$	0	0	0
$m_A(a_2) = 0.06$	0	$m(a_2) = 0.1223$	0	0
$m_A(a_3) = 0.35$	0	0	$m(a_3) = 0.6487$	0
$m_A(a_4) = 0.24$	0	0	0	$m(a_4) = 0.0667$
	$m_B(a_1) = 0.10$	$m_B(a_2) = 0.44$	$m_B(a_3) = 0.40$	$m_B(a_4) = 0.06$

In computing $P(H_i|E^A, E^B)$ in Eq. (4.30), the values for $P(H_i)$ are not specified by the example. Hence, they are set equal to each other for all i by the principle of indifference (also known as the insufficient reason principle) and do not play a role in the solution of the problem. For example, if $P(H_i)$ equal to 0.25 for all i were included, α would be 18.5357 (4 times larger), but the final values for $P(H_i|E^A, E^B)$ would be the same.

4.7 Constructing Probability Mass Functions

Perhaps the most difficult part of applying Dempster–Shafer theory in its original or modified forms is obtaining probability mass functions. Five methods for constructing these functions are presented next in no particular preference order.

4.7.1 Knowledge of sensor operation and object signature characteristics

The first procedure utilizes knowledge of the response of the sensors to the object of interest's area, shape, length, fabrication material, location, burial depth, etc., to create probability distributions that relate probability mass to the value of a parameter measured by the sensor. Milisavljević, Bloch, and Acheroy[9] and Milisavljević and Bloch[10,11] used this approach to detect antipersonnel mines buried in the soil. In part, their technique was driven by five considerations: (1) the pertinent data are images and sensor measurements, (2) data sets are not numerous enough to allow reliable statistical learning, (3) data are highly variable depending on the context and conditions, (4) data do not give precise information on the type of mine as there is ambiguity between several types, and (5) models are not available for every potential buried object.

Three sensors were deployed: an infrared (IR) camera, a metal detector (MD), and a ground-penetrating radar (GPR). Many experiments conducted with a particular type of IR camera revealed that the shape and area (elongation and ellipse fitting) of the camera images give information on the shape regularity of the detected object. The findings were twofold: (1) whenever the area is too small or too large, the object is not a mine and (2) whenever the area is within some range corresponding to the expected sizes of mines, the object can be a mine or anything else as well. Thus, the size of the image obtained from the IR camera is related to the belief that a regularly or irregularly shaped mine or a regularly or irregularly shaped nondangerous object is present. The probability mass functions that pertain to the elongation of the image were heuristic in nature, using information such as the ratio between minimum and maximum distances of bordering pixels measured from the center of gravity and the ratio of minor to major axes. Probability masses that relate to how well an image fits an ellipse were a function of object area, ellipse area, and object area within the ellipse area. Probability masses associated with image area were derived from an exponential function of the form $c \times \exp(N/D)$, where c is a function of the imaged object area and the lower limit for the approximate range of mine areas, N is a function of the imaged object area and the lower and upper limits for the approximate range of mine areas, and D is determined by the lower and upper limits for the mine areas.

The size of the metallic area in the MD data provides information on the shape, area, and burial depth of an object. This information assumes that the point spread function (impulse response) of the MD is known, data are not saturated, and the scanning step in both directions is sufficiently small. When these conditions are satisfied, the probability mass assignment by the MD is given in terms of an exponential function of the width of the scan.

For the GPR, the propagation velocity of the radar energy through the ground conveys information about material type or identity when burial depth data verify that the object is below the surface. In this circumstance, the propagation velocity should approximate that of the medium in which the object is buried. Burial depth of the object gives information concerning whether the object is a mine, as mines are expected only up to some maximum depth. Other objects can be found at any depth. The GPR also provides shape information as the ratio of object size to its scattering function as mine values are expected to lie within some known range. The probability mass functions deduced by propagation velocity and object shape information from the GPR are exponential in form, while the probability mass function for burial depth has a hyperbolic cosine relation.

The above method of probability mass assignment requires another assumption. The numerical representation of the probability mass functions presumes that we can assign numbers that represent degrees of belief. The general shapes and tendencies are derived from our knowledge of the types of information available from a particular sensor and the dependence of that information on sensor parameters and object shapes, areas, and locations. However, there remain some arbitrary choices that might appear as a limitation of the method, although it is not necessary to have precise estimates of the numbers, and ample robustness is observed when the values are used experimentally.

This can be explained in two ways. First, the representations for the probability mass functions are used for rough information. Hence, they do not have to be precise themselves. Second, several pieces of information are combined in the whole reasoning process, decreasing the influence of each particular value of individual information. Therefore, the chosen probability mass numbers are not crucial. What is important is the preservation of the ranking and shape of the functions that are derived from knowledge.[10]

4.7.2 Known probability distributions for the parameters of interest

The second method presumes that the probability distribution for the data of interest is known. Zhang, Hamedi, and Haghani[12] apply this technique to calculate probability mass values for travel-time windows from a cumulative distribution function that models travel time for a given time window as a lognormal distribution. The distribution is approximated using the estimated mean and sample variance of the data. Propositions belong to the high confidence band when they are located in proximity to the mean. Propositions belong to low confidence bands when they are located in the tails of the distribution. Therefore, propositions belonging to the high confidence band are more likely to reflect the actual travel time since they are close to the sample mean. The actual probability mass equals the difference in the cumulative distribution functions for the upper and lower bounds of the time window of interest multiplied by the credibility factor described in Section 2.13.4. The credibility factor weighs the contribution of real-time data and historical data to the cumulative distribution function.

4.7.3 Confusion matrix creation

The third method creates confusion matrices based on a comparison of real-time data from a sensor or another source with accurate ground-truth data. Entries along the diagonal of the confusion matrix are proportional to the number of instances when the sensor's real-time datum agrees with the true value of the measured parameter obtained from the ground-truth source. Off-diagonal entries correspond to the number of misclassified cases. The actual values of the probability masses are found by normalizing the confusion matrix using one of two strategies. The first assumes that the frame of discernment θ is included as a potential travel-time decision that models ignorance about the travel time on the part of the data source. In this strategy, normalization occurs by dividing each matrix element by the total of all of the matrix elements. In the second strategy, we are always able to select one of the travel-time intervals as the output of the data source. Here, normalization is performed by column; i.e., entries in each column of the confusion matrix are divided by the total of the column entries such that each column vector represents probability mass values that sum to unity. The second strategy is Bayesian in nature since no probability mass is assigned to the uncertainty class and all propositions (the travel-time intervals) are singletons.[13–15]

4.7.4 Number and degree of matching of features to those of objects of interest

A fourth method defines probability masses based on the number and ability of features extracted from incoming sensor data to match expected object traits.[16] Time-domain processing, for example, may use features such as signal amplitude, pulse width, amplitude-to-width ratio, rise and fall times, and pulse repetition frequency. Frequency domain processing may use separation between spectral peaks, widths of spectral features, identification of periodic structures in the signal, and number of scattering centers producing a return signal greater than a clutter-adaptive running-average threshold. Video detection systems may employ object discriminants such as image-fill criteria, where the number of pixels above some threshold is compared to the total number of pixels within the image boundaries, length-to-width ratio of the image (un-normalized or normalized to area or edge length), parallel and perpendicular line relationships, presence of arcs or circles or conic shapes in the image, central moments, center of gravity, asymmetry measures, and, with IR systems, temperature gradients across object boundaries. Multispectral and hyperspectral sensors operating in the visible and IR spectral bands may utilize color coefficients, apparent temperature, presence of specific spectral peaks or lines, and the space- and time-dependent signatures of the detected objects.

4.7.5 Exponential probability mass model

The fifth approach is credited to Denoeux.[8] The procedure was originally developed to assign a sample to one of M classes based on how close the sample was to several training vectors belonging to different classes. The Denoeux model expresses probability mass as a constant multiplied by an exponential function, where the exponent is proportional to the negative of the distance d (based on some distance measure, a table of which can be found in Hall[17]) between an input data sample to be classified and the kth nearest neighbors that belong to a specific class. Thus, the probability mass m is given by

$$m = \alpha_0 \exp[-\gamma_q d^{\beta}], \tag{4.31}$$

where $\gamma_q > 0$, q represents a class label from 1 .... M, where M is the number of classes in the set of classes C under consideration, and β is set equal to a small value (usually 1 or 2). In this problem, C is the frame of discernment θ. Denoeux reports that good results on average are obtained with $\alpha_0 = 0.95$ and γ_q determined separately for each class as $1/d_q^{\beta}$, where d_q is the mean distance between two training vectors belonging to class C_q. By substituting $\gamma_q = 1/d_q^{\beta}$ into Eq. (4.31), we obtain

$$m = \alpha_0 \exp\left[-\frac{d^{\beta}}{d_q^{\beta}}\right]. \tag{4.32}$$

The value of β was found to have very little influence on the performance of the method. Accordingly, a value of $\beta = 1$ was used by Denoeux. Kong, Li, Chen, and Liu[18] and Gao, Ozbay, and Kurkcu[19] modified the Denoeux probability mass model with a multiplicative dynamic reliability factor, as explained in Section 2.13.3.

4.8 Decision Support System Application of Dempster–Shafer Reasoning

Dempster–Shafer evidential reasoning was applied by Klein, Yi, and Teng[20] to create a decision support system at a traffic management center using a three-event, three-information source experiment. The system's goal was to enhance a traffic manager's understanding of the conditions giving rise to the data. The three potential events of interest that could occur at an intersection were

a_1 = downstream intersection congestion with spillback,
a_2 = one disabled vehicle blocking the right lane, and
a_3 = traffic flowing normally.

The information sources were normal roadside sensor data acquired by traffic operation centers (TOCs), cellular telephone messages, and citizen-band (CB) radio reports.

4.8.1 Field test description

Field testing of the Dempster–Shafer data fusion technique consisted of two parts. Part one included data acquisition and system calibration. Part two estimated the probability mass matrices required to generate the system output that was compared with the actual field event. A large data collection effort was conducted from March to mid-July 2001 in the Cleveland–Akron metropolitan area, which contains a population of two million people. Traffic incidents during the morning and afternoon peak periods were monitored on the city streets of Cleveland, Akron, and Cuyahoga Falls, and on major metropolitan freeways, including I-90, I-71, I-480, I-76, and I-77.

Surveillance cameras and system detectors from TOCs displayed traffic flow conditions on city streets and supported operations personnel in the rapid identification of problems. Incident reports from cellular telephones were obtained by monitoring commercial radio stations where many motorists call in to report congestion and incidents. The 911 dispatch center was not used as a data source due to privacy concerns and because only serious traffic incidents seem to generate emergency calls. Furthermore, calls to 911 represent a very small portion of the total telephone calls that are generated in response to traffic events. CB radio transmissions were continuously monitored to intercept traffic incident reports. The information on CB radio was primarily provided by commercial vehicles and pickup trucks.

Every detected traffic incident was subsequently verified by contacting the highway patrol or the city police. Some of the incidents could not be confirmed because they cleared quickly before the police could record any description of the event. Approximately 92 incidents were followed during the data collection period. Of these, 14 could not be verified and thus were not included in the database.

Table 4.9 lists the reporting rate and accuracy rate of the major field data sources. The reporting rate indicates the percent of the incidents conveyed by the various data sources. The accuracy rate specifies the percent of time each data source reported the same incident event as the highway patrol or city police. For example, a 70% accuracy rate for a CB radio report describing a vehicle blocking a lane indicates that such a report is correct 70% of the time. Although cellular telephones generated the largest number of incident reports, many of these reports were of poor quality, some even conflicting. In comparison, CB radio had a higher accuracy rate, but the number of reports was much smaller, in part, due to CB use by only a small number of vehicles. In addition, limited resources prevented monitoring of CB radio transmissions over the entire metropolitan area, further contributing to the low CB reporting rate. The accuracy rate for the TOC-reported events had the largest variation among the data sources. This rate can be much higher if additional surveillance equipment is deployed in the field to detect the events. Other data sources (TV, personal telephone calls) were found to have a high accuracy rate. However, this information arrived too late to be practically useful, and the number of these reports was very limited.

For every incident, incident reports were obtained from the data sources described above. Twelve of the 78 verified incidents were analyzed and categorized using Dempster's rule. A probability mass matrix derived

Table 4.9 Reporting rate and accuracy rate of data sources.

		Accuracy Rate (%)				
Data Source	**Reporting Rate (%)**	**Crash Rate**	**Vehicle Blocking One or More Lanes**	**Unanticipated Construction**	**Signal Error**	**Misc.[1]**
Cellular telephone	84	77	64	78	87	64
CB radio	5	82	70	81	90	66
TOC	8	80	78	63	95	57
Others[2]	3	89	n/a	85	n/a	n/a

[1]Material spills, animals on roadway, etc.
[2]TV, personal telephone calls (delayed).

from information from each data source was created each time an incident was reported. The matrix varied from incident to incident since different events were reported as causing the incident. Dempster's rule was used to identify the event with the highest probability as the most likely cause of the incident. It was found that the calibrated decision support system correctly identified 9 events (75% accuracy). In one of the three remaining events, the calculated probability of the true event was only slightly lower than that selected by the Dempster–Shafer data fusion process.

4.8.2 Field test conclusions

Using Dempster–Shafer theory to detect and identify events that affect highway traffic operations is conceptually clear and easy to apply. Testing demonstrates that the technique is effective in combining and processing reported traffic information from different data sources. The output of the process can offer timely information to assist in traffic management decision making since it is designed to extract whatever information exists in an otherwise incomplete database derived from a variety of data sources. Hence, it can assist in identifying a traffic incident type, guide the display of traffic advisory messages, and reduce congestion clearance times.

Field implementation of the Dempster–Shafer data fusion technique indicates that the most critical item affecting its effectiveness is the ability to calculate accurate probability mass values for each contributing data source. Traffic management personnel generated the probability masses from offline analyses of the acquired field data. Although this was only a feasibility study, our experience with the field data suggests that an operational test in a large metropolitan area is needed to quantify the dependencies of the probability masses associated with each data source on roadway type and configuration, and operating conditions such as traffic flow rate, time of day, day of week, seasonal variations, and weather. Test results show a large degree of variation in the reporting rates and accuracy rates of different data sources. The effectiveness of the decision support system depends on modifying the probability masses in response to changes in data sources and field conditions.

4.9 Comparison with Bayesian Inference

Dempster–Shafer evidential reasoning accepts an incomplete probabilistic model. Bayesian inference does not. Thus, Dempster–Shafer reasoning can be applied when the prior probabilities and likelihood functions or ratios are unknown. The available probabilistic information is interpreted as phenomena that impose truth values to various propositions for a certain time period, rather than as likelihood functions. Dempster–Shafer reasoning estimates how close the evidence is to forcing the truth of a hypothesis, rather than estimating how close the hypothesis is to being true.[21,22]

Dempster–Shafer reasoning allows sensor classification error to be represented by a probability assignment directly to an uncertainty class θ. Furthermore, Dempster–Shafer reasoning permits probabilities that express certainty or confidence to be assigned directly to an uncertain event, namely, any of the propositions in the frame of discernment θ or their unions. The Bayesian approach permits probabilities to be assigned only to the original propositions themselves. This is expressed mathematically in Bayesian inference as

$$P(a + b) = P(a) + P(b) \tag{4.33}$$

under the assumption that a and b are disjoint propositions. In Dempster–Shafer theory,

$$P(a + b) = P(a) + P(b) + P(a \cup b). \tag{4.34}$$

Shafer expresses the limitation of Bayesian inference in a more general way: "Bayesian theory cannot distinguish between lack of belief and disbelief. It does not allow one to withhold belief from a proposition without according that belief to the negation of the proposition."[4]

Bayesian theory does not have a convenient representation for ignorance or uncertainty. Prior distributions have to be known or assumed with Bayesian theory. A Bayesian support function ties all of its probability mass to single points in θ. There is no freedom of motion, i.e., no uncertainty interval.[23] The user of a Bayesian support function must somehow divide the support among singleton propositions. This may be easy in some situations such as an experiment with a fair die. If we believe that a fair die shows an even number, we can divide the support into three parts, namely, 2, 4, and 6. If the die is not fair, then Bayesian theory does not provide a solution.

Thus, the difficulty with Bayesian theory is in representing what we actually know without being forced to overcommit when we are ignorant. With Dempster–Shafer, we use information from the sensors (or other information sources) to find the support available for each proposition. For the fair-die example, Dempster–Shafer gives the probability mass m_k(even). If the die were not fair, Dempster–Shafer would still give the appropriate probability mass.

Therefore, there is no inherent difficulty in using Bayesian statistics when the required information is available. However, Dempster–Shafer reasoning offers an alternative approach when knowledge is not complete, e.g., when ignorance exists about the prior probabilities associated with the propositions in the frame of discernment, or when propositions are not mutually exclusive. The Dempster–Shafer formulation of a problem collapses into the Bayesian when the uncertainty interval is zero for all propositions and the probability mass assigned to unions of propositions is zero. On the other hand, any distinguishing proposition information that may have been available from prior probabilities is ignored when Dempster–Shafer in its original formulation is applied.

4.10 Modifications to the Original Dempster–Shafer Method

Criticism of Dempster–Shafer reasoning has been expressed concerning the way it reassigns probability mass originally allocated to conflicting propositions and the effect of the redistribution on the proposition selected as the output of the fusion process.[24,25] This concern is of particular consternation when there is a large amount of conflict that produces counterintuitive results.

A number of alternatives have been proposed to modify Dempster's rule to better accommodate conflicting beliefs.[5,16,26] Among these are the transferable belief model due to Smets[27,28] and Smets and Kennes,[29] which modifies the basic probability assignment in proportion to the number of elements it contains; the plausibility transformation due to Cobb and Shenoy,[30] which utilizes the concept of plausibility to decide among propositions; the modified Dempster–Shafer approach of Fixsen and Mahler,[31] which utilizes *a priori* probability measures as weighting functions on the probability masses; the methodology of Yager, which assigns conflicting probability mass to the uncertainty class (the frame of discernment) via the normalization factor;[32] and plausible and paradoxical reasoning due to Dezert,[33] which allows evidence from the conjunction (AND) operator $\cap$ as well as the disjunction (OR) operator $\cup$ to be admitted. The reader is referred to the above-cited references and to Klein[16] for further explanations of these methods.

The effect of conflicting propositions is illustrated by the following example attributed to Zadeh.[34] Consider two physicians who examine a patient and agree that the patient suffers from either meningitis M, concussion C, or brain tumor T. Thus, $\theta = \{M, C, T\}$. Furthermore, the doctors agree in their low expectation of a tumor, but disagree in the other likely cause and provide diagnoses as follows:

$$m_1(M) = 0.99 \quad m_1(T) = 0.01 \quad \text{and} \quad m_2(C) = 0.99 \quad m_2(T) = 0.01,$$

where the subscript 1 represents the diagnosis of physician 1, and subscript 2 represents the diagnosis of physician 2. Combining the physicians' belief functions using Dempster's rule to form the orthogonal sum gives

$$m(T) = 0.001/(1 - 0.0099 - 0.0099 - 0.9801) = 1. \tag{4.35}$$

This is certainly an unexpected result that arises from the bodies of evidence (the physicians' diagnoses) agreeing that the patient does not suffer from a tumor, but being in almost full contradiction for the other causes of the disease. The result conveys a negative implication for the use of the original Dempster–Shafer method in automated reasoning or data analysis where a counterintuitive result may not be uncovered. The reader can verify the result in Eq. (4.35) through an orthogonal sum calculation such as those shown in Tables 4.5 and 4.6. The issue posed by this example is solved by Dezert by allowing propositions in the frame of discernment to be constructed using the AND operator along with the original OR operator.[33]

The concern highlighted above is not unique to Dempster–Shafer. In general, the inherent uncertain nature of information sources, due to either sensor imprecision or changes in operator priorities, causes conditions to exist in which information fusion algorithms are inappropriate for those conditions.[35] Nevertheless, information acquisition and automated fusion of information may still be retained at a relatively high level as long as the operator has access to the raw data and is aware of the level of unreliability such that some attention will be allocated to the original data. These issues are also discussed in Section 2.12.

4.11 Summary

The Dempster–Shafer procedure allows multiple sensors or information sources to contribute evidence or data supporting a proposition as long as the sources are independent. Incomplete knowledge about propositions that corresponds to objects in a sensor's field of view is incorporated by assigning a portion of the sensor's probability mass to the uncertainty class. Dempster–Shafer can also assign probability mass to the union of propositions if the evidence supports it. It is in these regards that Dempster–Shafer differs from Bayesian inference, as Bayesian theory does not have a representation for uncertainty and permits probabilities to be assigned only to the original mutually exclusive propositions themselves.

The uncertainty interval is bounded on the lower end by the support for a proposition and on the upper end by the plausibility of the proposition. Support is the sum of *direct* sensor evidence for the proposition. Plausibility is the sum of all probability mass not directly assigned by the sensor to the negation of the proposition. Thus, the uncertainty interval depicts what proportion of evidence is truly in support of a proposition and what proportion results merely from ignorance.

Dempster's rule provides the formalism to combine probability masses from several sensors or information sources concerning the events of interest. If the intersections of the propositions form a null set, the probability masses of the null set elements are redistributed among the nonempty set members through a normalization factor. The intersection of propositions with the largest probability mass is usually selected as the output of the Dempster–Shafer fusion process. Another strategy for choosing the output is to select the proposition with the greatest plausibility.

Several alternative methods have been proposed to render the output of the Dempster–Shafer fusion process more intuitively appealing by preventing assignment of probability mass to highly conflicting propositions. Several of these approaches involve transformations of the support functions into probability functions that enable decision making based on the available information. An often-challenging component of applying Dempster–Shafer theory in its original or modified forms is obtaining the probability mass functions. The method to calculate these functions strongly depends on the experience and knowledge of the personnel assigned to this task and on known characteristics of the sensor output data.

An application of Dempster–Shafer reasoning to a decision support system that enhances a traffic manager's understanding of the conditions at the location from which the data are obtained was discussed. In this instance, the data of interest were obtained from roadway sensors, cellular telephone calls, and CB radios.

References

1. E. Waltz and J. Llinas, *Multisensor Data Fusion*, Artech House, Norwood, Massachusetts (1990).
2. R. A. Dillard, *Computing Confidences in Tactical Rule-Based Systems by Using Dempster–Shafer Theory*, NOSC TD 649 (AD A 137274), Naval Ocean Systems Center, San Diego, Sep. 14 (1983).
3. S. S. Blackman, *Multiple Target Tracking with Radar Applications*, Artech House, Norwood, Massachusetts (1986).
4. G. Shafer, *A Mathematical Theory of Evidence*, Princeton University Press, Princeton, (1976).
5. C. K. Murphy, "Combining belief functions when evidence conflicts," *Decision Support Systems*, **29**(1), 1–9 (2000).
6. P. L. Bogler, "Shafer–Dempster reasoning with applications to multisensor target identification systems," *IEEE Trans. on Systems, Man, and Cybernetics* **17**(6), 968–977 (1987).
7. T. D. Garvey, J. D. Lowrance, and M. A. Fischler, "An inference technique for integrating knowledge from disparate sources," *Proc. Seventh International Joint Conference on Artificial Intelligence* **I**, IJCAI-81, 319–325 (1981).
8. T. Denoeux, "A k-nearest neighbor classification rule based on Dempster–Shafer theory," *IEEE Trans. on Systems, Man, and Cybernetics* **25**(5), 804–813 (1995).
9. N. Milisavljević, I. Bloch, and M. Acheroy, "Characterization of mine detection sensors in terms of belief functions and their fusion, first

results," *IEEE Proc. Third International Conference on Information Fusion* **2**, Paris, July 10–13 (2000).

10. N. Milisavljević and I. Bloch, "Sensor fusion in anti-personnel mine detection using a two-level belief function model," *IEEE Trans. on Systems, Man, and Cybernetics–Part C: App. and Reviews* **33**(2), 269–283 (2003).
11. N. Milisavljević and I. Bloch, "Possibilistic versus belief function fusion for antipersonnel mine detection," *IEEE Trans. Geosci. and Rem. Sen.* **46**(5), 1488–1498 (2008).
12. X. Zhang, M. Hamedi, and A. Haghani, "A real-time data fusion framework for corridor travel time estimation with multiple data sources," Presented at 79th Annual Meeting of the Transportation Research Board, Paper 17-04974, Washington, D.C. (2017).
13. N.-E. El Faouzi, "Bayesian and evidential approaches for traffic data fusion: methodological issues and case study," Presented at 85th Annual Meeting of the Transportation Research Board, Paper 06-1510, Washington, D.C. (2006).
14. N.-E. El Faouzi, "Travel time estimation via evidential data fusion," *Recherche Transports Sécurité* **68**, 15–30 (2000).
15. N.-E. El Faouzi, L. A. Klein, and O. De Mouzon, "Improving travel time estimates from inductive loop and toll collection data with Dempster–Shafer data fusion," *Transportation Research Record, Journal of the Transportation Research Board, No. 2129*, Transportation Research Board, Washington, D.C., 73–80 (2009).
16. L. A. Klein, *Sensor and Data Fusion: A Tool for Information Assessment and Decision Making*, Second Edition, SPIE Press, Bellingham, Washington (2012) [doi: 10.1117.3.928035].
17. D. L. Hall, *Mathematical Techniques in Multisensor Data Fusion*, Artech House, Norwood, Massachusetts (1992).
18. Q.-J. Kong, Z. Li, Y. Chen, and Y. Liu, "An approach to urban traffic state estimation by fusing multisource information," *IEEE Trans. on Intelligent Transportation Systems* **10**(3), 499–511 (2009).
19. J. Gao, K. Ozbay, and A. Kurkcu, "Multi-source data fusion for urban traffic state estimation: a case study of New York City," Presented at the 98th Annual Meeting of the Transportation Research Board, Paper 19-04253, Washington, D.C. (2019).
20. L. A. Klein, P. Yi, and H. Teng, "Decision support system for advanced traffic management through data fusion," in *Transportation Research Record, Journal of the Transportation Research Board, No. 1804: Transportation Data and Information Technology Research*, Transportation Research Board, Washington, D.C., 173–178 (2002).
21. J. Pearl, *Probabilistic Reasoning in Intelligent Systems: Networks of Plausible Inference*, Morgan Kaufmann Publishers, San Mateo, California (1988).

22. D. Koks and S. Challa, "An introduction to Bayesian and Dempster–Shafer data fusion," DSTO-TR-1436, DSTO Systems Sciences Laboratory, Australian Government Department of Defence (2005).
23. J. A. Barnett, "Computational methods for a mathematical theory of evidence," *Proc. Seventh International Joint Conference on Artificial Intelligence* **II**, IJCAI-81, 868–875 (1981).
24. L. A. Zadeh, "A theory of approximate reasoning," in *Machine Intelligence* **9**, J. Hayes, D. Michie, and L. Mikulich, Eds., Elsevier, New York, pp. 149–194 (1979).
25. L. A. Zadeh, "Review of Shafer's A Mathematical Theory of Evidence," *AI Magazine* **5**(3), 81–83 (1984).
26. L. Valet, G. Mauris, and Ph. Bolon, "A statistical overview of recent literature in information fusion," *IEEE Aerospace and Electronic Systems Magazine* **16**(3), 7–14 (2001).
27. P. Smets, "Constructing the pignistic probability function in a context of uncertainty," *Machine Intelligence and Pattern Recognition* **10**, 29–39 (1990).
28. P. Smets, "The transferable belief model and random sets," *International Journal of Intelligent Systems* **7**(1), 37–46 (1992).
29. P. Smets and R. Kennes, "The transferable belief model," *Artificial Intelligence* **66**(2), 191–234 (1994).
30. B. R. Cobb and P. P. Shenoy, "A comparison of methods for transforming belief function models to probability models," *Lecture Notes in Artificial Intelligence*, in *Symbolic and Quantitative Approaches to Reasoning with Uncertainty*, T. D. Nielsen and N. L. Zhang, Eds., Springer-Verlag, Berlin, pp. 255–266 (2003).
31. D. Fixsen and R. P. S. Mahler, "The modified Dempster–Shafer approach to classification," *IEEE Trans. on Systems, Man, and Cybernetics–Part A: Systems and Humans* **27**(1), 96–104 (1997).
32. R. R. Yager, "On the Dempster–Shafer framework and new combination rules," *Information Science* **41**(2), 93–137 (1987).
33. J. Dezert, "Foundations for a new theory of plausible and paradoxical reasoning," *Information and Security* **9**, 13–57 (2002).
34. L. A. Zadeh, "On the Validity of Dempster's Rule of Combinations of Evidence," ERL Memo M 97/24, University of California, Berkeley (1979).
35. R. Parasuraman, T. B. Sheridan, and C. D. Wickens, "A model for types and levels of human interaction with automation," *IEEE Trans. on Systems, Man, and Cybernetics–Part A: Systems and Humans* **30**(3), 286–297 (2000).

Chapter 5
Kalman Filtering for Traffic Management

The Kalman filter provides a general solution to the recursive, minimum mean-square state estimation problem within the class of linear estimators, assuming that the dynamics of the object of interest and measurement noise are accurately modeled. As applied to scenarios where a radar or another active sensor is used to track objects, this filter estimates the object's state (e.g., position and velocity) at some time, usually the predicted time of the next observation, and then updates that estimate using noisy measurements. It also offers an estimate of the object's tracking error statistics through the state error-covariance matrix.[1–16]

As part of the optimal state estimation process, the Kalman filter calculates a filter gain that is dependent on the assumed object motion and measurement noise models. The gain can be used to define a gate test statistic that assists in correlating new measurements with existing tracks or in forming new tracks based on several successive measurements. The Kalman filter discussion presented in this chapter was originally developed for defense applications such as radar tracking of potential threat targets. However, the same techniques are directly applicable to the tracking of motorized and nonmotorized vehicles, and pedestrians. A time-invariant filter known as an α-β filter is also described for estimating the state of an object moving with constant velocity. The chapter concludes with descriptions of an interacting multiple model that can be used to track vehicles as they maneuver on a roadway and the extended Kalman filter, which often finds applications in estimating the state of nonlinear systems.

5.1 Optimal Estimation

According to Gelb et al.,[1] "an optimal estimator is a computational algorithm that processes measurements to deduce a minimum error (in accordance with some stated criterion of optimality) estimate of the state of a system by utilizing: knowledge of system and measurement dynamics, assumed statistics of system noises and measurement errors, and initial condition information." Thus, the optimal estimator minimizes the estimation error in a well-defined statistical sense and utilizes all available measurement data and prior knowledge concerning the system. Its limitations include its sensitivity to erroneous *a priori* models and statistics, and a potential computational burden (that may not be significant when using present-day, fast, low-cost microprocessors to estimate vehicle and pedestrian movements).

There are three types of estimation problems, namely, filtering, smoothing, and prediction. Filtering applies when the time at which an estimate is desired coincides with the time of the last measurement opportunity. Smoothing occurs when the time of interest lies between the time interval spanning the available measurement data. Finally, prediction occurs when an estimate is needed for a time falling after the last available measurement. The Kalman filter performs filtering and prediction.

5.2 Kalman Filter Application to Object Tracking

Figure 5.1 illustrates the application of the Kalman filtering process to the estimation of an object's state when an external control function is present, such as with an accelerating or turning vehicle. Measuring devices provide the values of pertinent observable system parameters at discrete time increments. These measurements contain all of the information that is explicitly available from the physical system for estimating its state. Often the state variables of interest cannot be measured directly, and some means of inferring their values from the measurements is needed. For example, connected vehicles may provide data that give the velocity of the vehicle or data from which the velocity can be inferred. These data may not be available from conventional vehicles or vehicles that do not incorporate dedicated short-range communications (DSRC) or some other form of standardized communications. However, if it is known that the conventional vehicle is a member of a platoon or similar group of which the connected vehicle is a member, it can be reasonably assumed that the former vehicle is traveling at the same speed as the latter.

According to Bar-Shalom and Fortmann, a state is loosely defined as the vector of smallest dimension that summarizes the past history of the system sufficiently to predict its future trajectory, assuming that future inputs will be

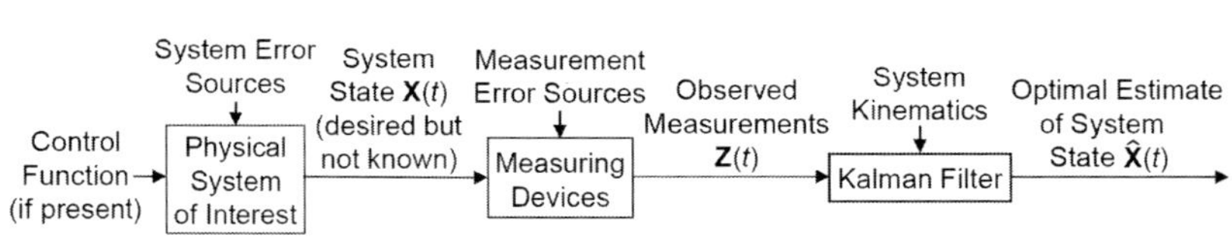

Figure 5.1 Kalman filter application to optimal estimation of the state of a linear system.

available.[7] For linear motion of an object in Cartesian coordinates, state space $\mathbf{X}$ is expressed by Eq. (5.1) when sensors provide measurements in three dimensions (e.g., 3D radars), where $\dim(\mathbf{X}) = 6$ such that

$$\mathbf{X}^{\mathrm{T}} = [x \quad y \quad z \quad \mathrm{d}x \quad \mathrm{d}y \quad \mathrm{d}z], \tag{5.1}$$

and by Eq. (5.2) when sensors provide measurements in two dimensions (e.g., 2D radars), where $\dim(\mathbf{X}) = 4$ such that

$$\mathbf{X}^{\mathrm{T}} = [x \quad y \quad \mathrm{d}x \quad \mathrm{d}y], \tag{5.2}$$

and where the superscript T indicates the transpose operation. Three-dimensional radars measure range, range rate, azimuth, and elevation or height, while 2D radars measure range and azimuth.

Since Kalman filtering predicts the state estimate and state error covariance and then updates them based on noisy measurements, we next define the discrete-time state-transition model and measurement model used in these processes. The models also provide an estimate of object-tracking error statistics through the state error-covariance matrix.

5.3 State Transition Model

The Kalman filter addresses the general problem of estimating the state $\mathbf{X} \in \Re^{n_x}$ of a discrete-time dynamic process governed by the linear stochastic difference equation

$$\mathbf{X}_{k+1} = \mathbf{F}\mathbf{X}_k + \mathbf{J}\mathbf{u}_k + \mathbf{w}_k \tag{5.3}$$

with a measurement $\mathbf{Z} \in \Re^{n_z}$, where $\mathbf{Z}$ is of dimension n_z. The measurement model is discussed in the next section. The state of the object of interest at time t_{k+1} is represented by $\mathbf{X}_{k+1}$ of dimension n_x; $\mathbf{u}_k$ is the known input driving or control function of dimension n_u; $\mathbf{F}$ is the known $n_x \times n_x$ state transition matrix or fundamental matrix of the system (sometimes denoted by $\mathbf{\Phi}$), here assumed to be independent of time but may not be in general; $\mathbf{J}$ is the $n_x \times n_u$ input matrix that relates $\mathbf{u}_k$ at the previous time step to the state at the current time; and $\mathbf{w}_k$ is the white process or plant noise having a zero-mean normal probability distribution

$$p(\mathbf{w}_k) \sim \mathrm{N}[0, \mathbf{Q}_k] \tag{5.4}$$

such that the expected value of its mean is

$$\mathrm{E}[\mathbf{w}_k] = 0, \tag{5.5}$$

the expected value of its covariance is

$$\mathrm{E}[\mathbf{w}_k \mathbf{w}_j^{\mathrm{T}}] = \begin{cases} \mathbf{Q}_k & \text{if } j = k \\ 0 & \text{otherwise} \end{cases}, \tag{5.6}$$

and $\mathbf{Q}_k$ is the matrix of the covariance values of $\mathbf{w}_k$ at time t_k.

Section 5.11.1 shows the state transition matrix to be of the form

$$\mathbf{F} = \mathrm{e}^{\mathbf{A}\Delta T} = \begin{bmatrix} 1 & \Delta T \\ 0 & 1 \end{bmatrix} \tag{5.7}$$

for a constant-velocity object, i.e., one for which acceleration is nominally zero. Denoting x as a generic coordinate allows the state vector to be written as $\mathbf{X}^{\mathrm{T}} = [x \quad \dot{x}]$, where the dot over the x indicates differentiation with respect to time, and ΔT is the time interval between samples, i.e., $t_{k+1} - t_k$.

Given a corrected (also referred to as an updated or filtered) state estimate $\hat{\mathbf{X}}_{k|k}$ at time t_k, the predicted state $\hat{\mathbf{X}}_{k+1|k}$ at time t_{k+1} can be expressed as

$$\hat{\mathbf{X}}_{k+1|k} = \mathbf{F}\hat{\mathbf{X}}_{k|k} \tag{5.8}$$

and the state error-covariance matrix $\mathbf{P}_{k+1|k}$ for the predicted state $\hat{\mathbf{X}}_{k+1|k}$ as

$$\mathbf{P}_{k+1|k} = \mathbf{F}\mathbf{P}_{k|k}\mathbf{F}^{\mathrm{T}} + \mathbf{Q}_k, \tag{5.9}$$

where $\mathbf{P}_{k|k}$ is the error-covariance matrix for the corrected state estimate at time t_k, and the notation $k+1|k$ indicates the predicted value (also referred to as the estimated or extrapolated value) at time $k+1$ calculated with data gathered at time k.

5.4 Measurement Model

The measurement is associated with the state through an equation of the form

$$\mathbf{Z}_k = \mathbf{H}\mathbf{X}_k + \boldsymbol{\beta}_k + \boldsymbol{\varepsilon}_k, \tag{5.10}$$

where $\mathbf{Z}_k$ represents the sensor measurement parameters at time t_k, $\mathbf{H}$ is the $n_z \times n_x$ observation matrix that relates the state to the measurement, $\mathbf{X}_k$ is the object state at time t_k, $\boldsymbol{\beta}_k$ is a fixed but unknown measurement bias error, and $\boldsymbol{\varepsilon}_k$ is the random component of the measurement error characterized as white noise having a zero-mean normal probability distribution

$$p(\boldsymbol{\varepsilon}_k) \sim \mathrm{N}[0, \mathbf{R}_k] \tag{5.11}$$

such that

$$\mathrm{E}[\boldsymbol{\varepsilon}_k] = 0, \tag{5.12}$$

$$\mathrm{E}[\boldsymbol{\varepsilon}_k \boldsymbol{\varepsilon}_j^{\mathrm{T}}] = \begin{cases} \mathbf{R}_k & \text{if } j = k \\ 0 & \text{otherwise} \end{cases}, \tag{5.13}$$

and $\mathbf{R}_k$ is the matrix of the covariance values of $\boldsymbol{\varepsilon}_k$ at time t_k. The bias error $\boldsymbol{\beta}$ is typically accounted for as part of sensor registration. Sensor registration deals with aligning multiple sensors to a common coordinate system. Good registration is required for satisfactory track initiation and measurement-to-track correlation. Improved registration also reduces the requirement for man–machine interfaces needed to resolve track initiation and correlation errors.

Since the bias error is dispensed with as part of the registration procedure, only the random error $\boldsymbol{\varepsilon}$ is retained in Eq. (5.10) such that

$$\mathbf{Z}_k = \mathbf{H}\mathbf{X}_k + \boldsymbol{\varepsilon}_k. \tag{5.14}$$

The process and measurement noise are usually assumed to be uncorrelated. Thus,

$$\mathrm{E}[\mathbf{w}_k \boldsymbol{\varepsilon}_j^{\mathrm{T}}] = 0 \text{ for all } j \text{ and } k. \tag{5.15}$$

5.4.1 Measurement error-covariance matrix for a 3D and 2D problem

When a 3D radar is used for measurements, $\mathbf{X}^{\mathrm{T}}$ is $[x\ y\ z\ \mathrm{d}x\ \mathrm{d}y\ \mathrm{d}z]$, and $\mathbf{Z}^{\mathrm{T}}$ is $[x\ y\ z]$. In this case,

$$\mathbf{H} = [\mathbf{I} \quad \mathbf{0}] = \begin{bmatrix} 1 & 0 & 0 & 0 & 0 & 0 \\ 0 & 1 & 0 & 0 & 0 & 0 \\ 0 & 0 & 1 & 0 & 0 & 0 \end{bmatrix}, \tag{5.16}$$

where $\mathbf{I}$ and $\mathbf{0}$ are the 3×3 identity matrix and 3×3 null matrix, respectively. The measurement error-covariance matrix $\mathbf{R}_k$ has the general form

$$\mathbf{R}_k = \begin{bmatrix} \sigma_{xx} & \sigma_{xy} & \sigma_{xz} \\ \sigma_{xy} & \sigma_{yy} & \sigma_{yz} \\ \sigma_{xz} & \sigma_{yz} & \sigma_{zz} \end{bmatrix}. \tag{5.17}$$

Transportation and traffic management applications are more often concerned with 2D measurements involving the x and y spatial coordinates. For this situation, $\mathbf{X}^{\mathrm{T}}$ is $[x\ y\ \mathrm{d}x\ \mathrm{d}y]$, $\mathbf{Z}^{\mathrm{T}}$ is $[x\ y]$, and

$$\mathbf{H} = [\mathbf{I} \quad \mathbf{0}] = \begin{bmatrix} 1 & 0 & 0 & 0 \\ 0 & 1 & 0 & 0 \end{bmatrix}, \tag{5.18}$$

where **I** and **0** are the 2×2 identity matrix and 2×2 null matrix, respectively, and the measurement error-covariance matrix $\mathbf{R}_k$ is expressed as

$$\mathbf{R}_k = \begin{bmatrix} \sigma_{xx} & \sigma_{xy} \\ \sigma_{xy} & \sigma_{yy} \end{bmatrix}. \tag{5.19}$$

5.4.2 Object in straight-line motion

Suppose that we wish to estimate the state of an object moving along a straight line at constant speed with a set of discrete-time measurements of its position. If the set of measurements $\mathbf{Z}_k$ is denoted by $\{Z_0, Z_1, Z_2, \ldots, Z_{M-1}\}$, the relation of the measurements to the initial position x_0 and speed v_0 is given by

$$\mathbf{Z}_k = x_0 + k(\Delta T)v_0 + \boldsymbol{\varepsilon}_k \text{ for } k = 0,\ 1,\ \ldots,\ (M-1), \tag{5.20}$$

where k represents the measurement number, ΔT is the sampling interval, and $\boldsymbol{\varepsilon}_k$ is the random component of the measurement error for the kth measurement. Assuming that the measurement errors have zero mean and standard deviation σ_k, the expected value of $\boldsymbol{\varepsilon}_k$ is zero, as given by Eq. (5.12), and the expected value of $\boldsymbol{\varepsilon}_k^2$ is

$$\mathrm{E}[\boldsymbol{\varepsilon}_k^2] = \mathbf{R}_k = \sigma_k^2 \text{ for } k = 0,\ 1,\ \ldots,\ (M-1). \tag{5.21}$$

The M scalar equations represented by Eq. (5.20) are written more compactly in matrix-vector form [as was shown in Eq. (5.14)] as $\mathbf{Z}_k = \mathbf{H}_k\mathbf{X} + \boldsymbol{\varepsilon}_k$, where

$$\mathbf{H}_k = \begin{bmatrix} 1 & 0 \\ 1 & \Delta T \\ 1 & 2(\Delta T) \\ \vdots & \vdots \\ 1 & (M-1)\Delta T \end{bmatrix} \tag{5.22}$$

is the observation matrix,

$$\mathbf{X} = \begin{bmatrix} x_0 \\ v_0 \end{bmatrix} \tag{5.23}$$

is the state whose estimate is to be updated by the measurements, and

$$\mathbf{R}_k = \begin{bmatrix} \sigma_0^2 \\ \sigma_1^2 \\ \vdots \\ \sigma_{M-1}^2 \end{bmatrix} \tag{5.24}$$

is the measurement error-covariance matrix.

For verification, we substitute Eqs. (5.22) through (5.24) into Eq. (5.14) to recover the M scalar equations of Eq. (5.20) as

$$\begin{bmatrix} Z_0 \\ Z_1 \\ Z_2 \\ \vdots \\ Z_{M-1} \end{bmatrix} = \begin{bmatrix} 1 & 0 \\ 1 & \Delta T \\ 1 & 2(\Delta T) \\ \vdots & \vdots \\ 1 & (M-1)\Delta T \end{bmatrix} \begin{bmatrix} x_0 \\ v_0 \end{bmatrix} + \begin{bmatrix} \sigma_0^2 \\ \sigma_1^2 \\ \sigma_2^2 \\ \vdots \\ \sigma_{M-1}^2 \end{bmatrix}. \tag{5.25}$$

5.5 The Discrete-Time Kalman Filter Algorithm

In the following discussion, the sampling intervals ΔT are constant. Therefore, $\mathbf{F}$, $\mathbf{J}$, and $\mathbf{H}$ do not depend on k. Also, $\mathbf{w}_k$ and $\boldsymbol{\varepsilon}_k$ are assumed to be constant, i.e., independent of time step k. Thus, $\mathbf{Q}$ and $\mathbf{R}$ are independent of k, and the discrete-time system is completely time invariant.

The Kalman filter computes a corrected, i.e., an updated, filtered, or *a posteriori*, state estimate $\hat{\mathbf{X}}_{k+1|k+1}$ at time step $k+1$ given a measurement $\mathbf{Z}_{k+1}$, which is expressed as a linear combination of a predicted or *a priori* estimate $\hat{\mathbf{X}}_{k+1|k}$ and a weighted difference between the actual measurement $\mathbf{Z}_{k+1}$ and a measurement prediction $\mathbf{H}\hat{\mathbf{X}}_{k+1|k}$. Algebraically, the corrected state estimate is written as

$$\hat{\mathbf{X}}_{k+1|k+1} = \hat{\mathbf{X}}_{k+1|k} + \mathbf{G}_{k+1}\,(\mathbf{Z}_{k+1} - \mathbf{H}\hat{\mathbf{X}}_{k+1|k}), \tag{5.26}$$

where the predicted estimate $\hat{\mathbf{X}}_{k+1|k}$ is given by Eq. (5.40) or (5.41). The $n_x \times n_z$ Kalman gain matrix $\mathbf{G}_{k+1}$ (assumed to be constant throughout a sampling interval) is selected to minimize the scalar sum of the diagonal elements, i.e., the trace, of the corrected covariance matrix $\mathbf{P}_{k+1|k+1}$ at time $k+1$, where[1]

$$\mathbf{P}_{k+1|k+1} = \mathrm{E}[(\mathbf{X}_{k+1} - \hat{\mathbf{X}}_{k+1|k+1})(\mathbf{X}_{k+1} - \hat{\mathbf{X}}_{k+1|k+1})^{\mathrm{T}}]. \tag{5.27}$$

That value of $\mathbf{G}_{k+1}$ is

$$\mathbf{G}_{k+1} = \mathbf{P}_{k+1|k}\,\mathbf{H}^{\mathrm{T}}\,(\mathbf{H}\mathbf{P}_{k+1|k}\,\mathbf{H}^{\mathrm{T}} + \mathbf{R})^{-1}, \tag{5.28}$$

where dim($\mathbf{G}$) is $2n_z \times n_z$.

The diagonal elements of the covariance matrix are the mean-square errors in knowledge of the state variables. The off-diagonal elements indicate the cross-correlation between the state variables, typically x and v. These terms are related to the linear correlation coefficient $\rho(x, v)$ by[1]

$$\rho(x,\, v) = \frac{\mathrm{E}[xv]}{\sigma_x \sigma_v}, \tag{5.29}$$

where σ represents the standard deviation.

The difference $(\mathbf{Z}_{k+1} - \mathbf{H}\hat{\mathbf{X}}_{k+1|k})$ appearing in Eq. (5.26) is the measurement innovation or residual. A residual of zero implies complete agreement between the measurement and prediction. The second term of the measurement innovation $\mathbf{H}\hat{\mathbf{X}}_{k+1|k}$ is referred to as the measurement prediction $\hat{\mathbf{Z}}_{k+1|k}$.

Process noise $\mathbf{w}_{k+1}$ is defined as the difference between the actual value of the measurement and its predicted value or, equivalently, as the innovation. Thus,

$$\mathbf{w}_{k+1} \equiv \mathbf{Z}_{k+1} - \hat{\mathbf{Z}}_{k+1|k} \tag{5.30}$$

$$= \mathbf{Z}_{k+1} - \mathbf{H}\hat{\mathbf{X}}_{k+1|k}. \tag{5.31}$$

The covariance matrix $\mathbf{S}_{k+1}$ of the innovation equals

$$\mathbf{S}_{k+1} = \mathrm{cov}[\mathbf{Z}_{k+1} - \mathbf{H}\hat{\mathbf{X}}_{k+1|k}] = \mathbf{H}\mathbf{P}_{k+1|k}\mathbf{H}^{\mathrm{T}} + \mathbf{R}. \tag{5.32}$$

The corrected error-covariance matrix $\mathbf{P}_{k+1|k+1}$ may be written in several forms that follow from its definition in Eq. (5.27).[1,7,8] Accordingly,

$$\mathbf{P}_{k+1|k+1} = \mathbf{P}_{k+1|k} - \mathbf{G}_{k+1}\,\mathbf{S}_{k+1}(\mathbf{G}_{k+1})^{\mathrm{T}} \tag{5.33}$$

$$= (\mathbf{I} - \mathbf{G}_{k+1}\,\mathbf{H})\mathbf{P}_{k+1|k}(\mathbf{I} - \mathbf{G}_{k+1}\,\mathbf{H})^{\mathrm{T}} + \mathbf{G}_{k+1}\,\mathbf{R}(\mathbf{G}_{k+1})^{\mathrm{T}} \tag{5.34}$$

$$= (\mathbf{I} - \mathbf{G}_{k+1}\,\mathbf{H})\mathbf{P}_{k+1|k}, \tag{5.35}$$

where $\mathbf{I}$ is the identity matrix.

The different structures for the $\mathbf{P}_{k+1|k+1}$ covariance equations have different numerical properties. For example, at the expense of some extra computation, the quadratic form of Eq. (5.34) guarantees that $\mathbf{P}_{k+1|k}$ and $\mathbf{R}$ will remain symmetric and $\mathbf{P}_{k+1|k+1}$ positive definite. The form of $\mathbf{P}_{k+1|k+1}$ in Eq. (5.35) is used to calculate the Kalman gain.

Incorporating the object dynamics and measurement models from Eqs. (5.3) and (5.14) gives the set of Kalman filter equations as

$$\mathbf{G}_{k+1} = \mathbf{P}_{k+1|k}\mathbf{H}^{\mathrm{T}}(\mathbf{H}\mathbf{P}_{k+1|k}\mathbf{H}^{\mathrm{T}} + \mathbf{R})^{-1} = \mathbf{P}_{k+1|k}\,\mathbf{H}^{\mathrm{T}}(\mathbf{S}_{k+1})^{-1}, \tag{5.36}$$

$$\hat{\mathbf{X}}_{k+1|k+1} = \hat{\mathbf{X}}_{k+1|k} + \mathbf{G}_{k+1}(\mathbf{Z}_{k+1} - \mathbf{H}\hat{\mathbf{X}}_{k+1|k}) \tag{5.37}$$

$$= [\mathbf{I} - \mathbf{G}_{k+1}\mathbf{H}]\hat{\mathbf{X}}_{k+1|k} + \mathbf{G}_{k+1}\mathbf{Z}_{k+1}, \tag{5.38}$$

$$\mathbf{P}_{k+1|k+1} = (\mathbf{I} - \mathbf{G}_{k+1}\mathbf{H})\,\mathbf{P}_{k+1|k}, \tag{5.39}$$

$$\hat{\mathbf{X}}_{k+1|k} = \mathbf{F}\hat{\mathbf{X}}_{k|k} + \mathbf{J}\mathbf{u}_k \text{ when a driving or control function is present, or} \tag{5.40}$$

$$\hat{\mathbf{X}}_{k+1|k} = \mathbf{F}\hat{\mathbf{X}}_{k|k} \text{ in the absence of a driving or control function, and} \tag{5.41}$$

$$\mathbf{P}_{k+1|k} = \mathbf{F}\mathbf{P}_{k|k}\,\mathbf{F}^{\mathrm{T}} + \mathbf{Q}. \tag{5.42}$$

An alternate expression for the Kalman gain is[1,7]

$$\mathbf{G}_k = \mathbf{P}_{k|k}\mathbf{H}^{\mathrm{T}}\mathbf{R}^{-1}. \tag{5.43}$$

Thus, the gain matrix is proportional to the uncertainty in the state estimate and inversely proportional to the measurement noise. If measurement noise is large and state estimate errors are small, the measurement is due primarily to the noise, and only small changes in the state estimates should be made. This is precisely the result given by Eq. (5.36) or (5.43). On the other hand, small measurement noise and large uncertainty in the state estimates suggest that the measurement contains considerable information about errors in the estimates. Therefore, the difference between the actual and the predicted measurement forms the basis for making strong corrections to the estimates. This consequence is indicated clearly by Eq. (5.43). Hence, the filter gain matrix calculation agrees with an intuitive approach to improving the estimate.[1]

Equation (5.41) describes the state transition when vehicles are traveling at constant velocity. Equation (5.40) is applicable when vehicles are accelerating or changing direction, e.g., turning a corner or changing lanes.

Blackman observes that a version of the Kalman filter may be defined in which the filtered quantities (i.e., $\hat{\mathbf{X}}_{k+1|k+1}$ and $\mathbf{P}_{k+1|k+1}$) are bypassed, and only one-step-ahead prediction quantities ($\hat{\mathbf{X}}_{k+1|k}$ and $\mathbf{P}_{k+1|k}$) are used.[3] This is important for real-time operation of multiple-object tracking systems where often only predicted quantities are of practical importance. In this formulation, the pertinent equations are

$$\mathbf{G}_k = \mathbf{P}_{k|k-1}\mathbf{H}^{\mathrm{T}}(\mathbf{H}\mathbf{P}_{k|k-1}\mathbf{H}^{\mathrm{T}} + \mathbf{R})^{-1}, \tag{5.44}$$

$$\hat{\mathbf{X}}_{k+1|k} = \mathbf{F}[\hat{\mathbf{X}}_{k|k-1} + \mathbf{G}_k(\mathbf{Z}_k - \mathbf{H}\hat{\mathbf{X}}_{k|k-1})], \tag{5.45}$$

$$\mathbf{P}_{k+1|k} = \mathbf{F}[(\mathbf{I} - \mathbf{G}_k\,\mathbf{H})\mathbf{P}_{k|k-1}]\mathbf{F}^{\mathrm{T}} + \mathbf{Q}. \tag{5.46}$$

Equation (5.45) is obtained by substituting Eq. (5.37) into Eq. (5.41), while Eq. (5.46) is found by substituting Eq. (5.39) into Eq. (5.42). Equation (5.46) may be written in other forms by replacing the gain factor with its equivalent representation from Eq. (5.44).

Figure 5.2 separates the Kalman filter equations into two clusters: those that predict the state at the time of the next update and those that correct the state prediction using measurement updates. The prediction equations [Eq. (5.40) or Eqs. (5.41) and (5.42)] project forward the estimates of the current state and error-covariance values to obtain the *a priori* estimates for the next time step. The correction equations [Eqs. (5.36), (5.37), and (5.39)] incorporate feedback of noisy measurements into the *a priori* state estimate to obtain the filtered or improved *a posteriori* state estimate. In the radar tracking application, the correction equations adjust the predicted track estimate based on the actual measurement at that time.[5] At every measurement k, the entire past is summarized by the sufficient statistic $\hat{\mathbf{X}}_{k|k}$ and its associated covariance $\mathbf{P}_{k|k}$.

A summary of the Kalman filter state prediction and correction procedures, and of pertinent equations by Bar-Shalom and Fortmann[7] and Bar-Shalom, Willett, and Tian[8] is shown in Fig. 5.3. These authors divide the process into four major parts: evolution of the system, control function

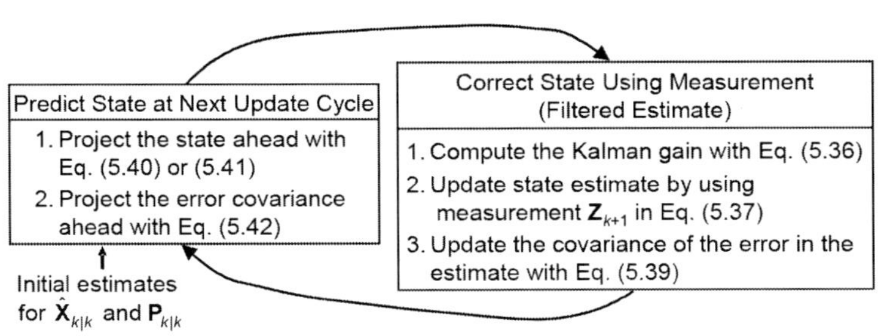

Figure 5.2 Discrete Kalman filter recursive operation.

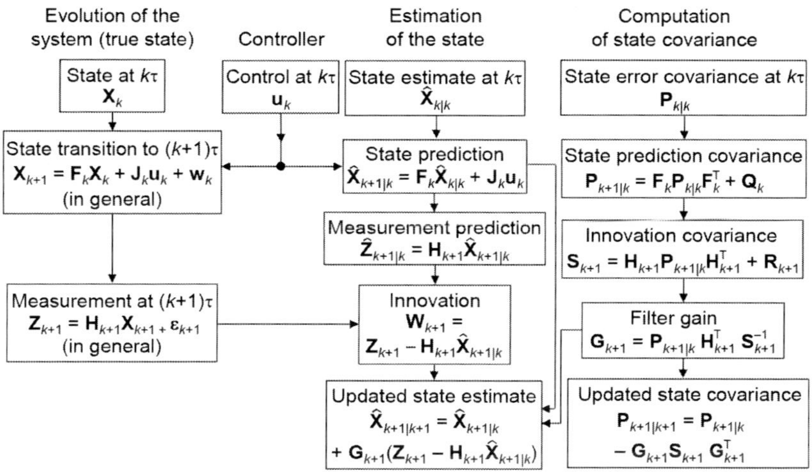

Figure 5.3 Kalman filter update process (adapted from Ref. 8 with permission).

activation, estimation of the state, and computation of the state error covariance. The subscript k on the state transition, control, and observation matrices, and on the process and measurement noise terms indicates that they can be time dependent in general.

State prediction and correction equations are linear since the state correction is a linear combination of prediction and measurement. If the measurement errors are normally distributed, then the predicted and corrected states are also normally distributed random variables. Empirical radar target tracking data suggest that the measurement errors in the Cartesian plane are normally distributed, at least approximately. Since there are many error sources within the radar hardware and software, the central limit theorem would seem to confirm this conclusion. However, empirical errors are not zero mean.[17]

The error-covariance matrix $\mathbf{P}$ for 3D measurement data has the general form

$$\mathbf{P} = \begin{bmatrix} \sigma_{xx} & \sigma_{xy} & \sigma_{xz} & \sigma_{x\dot{x}} & \sigma_{x\dot{y}} & \sigma_{x\dot{z}} \\ \sigma_{yx} & \sigma_{yy} & \sigma_{yz} & \sigma_{y\dot{x}} & \sigma_{y\dot{y}} & \sigma_{y\dot{z}} \\ \sigma_{zx} & \sigma_{zy} & \sigma_{zz} & \sigma_{z\dot{x}} & \sigma_{z\dot{y}} & \sigma_{z\dot{z}} \\ \sigma_{x\dot{x}} & \sigma_{x\dot{y}} & \sigma_{x\dot{z}} & \sigma_{\dot{x}\dot{x}} & \sigma_{\dot{x}\dot{y}} & \sigma_{\dot{x}\dot{z}} \\ \sigma_{y\dot{x}} & \sigma_{y\dot{y}} & \sigma_{y\dot{z}} & \sigma_{\dot{y}\dot{x}} & \sigma_{\dot{y}\dot{y}} & \sigma_{\dot{y}\dot{z}} \\ \sigma_{z\dot{x}} & \sigma_{z\dot{y}} & \sigma_{z\dot{z}} & \sigma_{\dot{z}\dot{x}} & \sigma_{\dot{z}\dot{y}} & \sigma_{\dot{z}\dot{z}} \end{bmatrix} = \begin{bmatrix} \mathbf{P}_p & \mathbf{P}_c \\ \mathbf{P}_c^{\mathrm{T}} & \mathbf{P}_v \end{bmatrix}, \tag{5.47}$$

where the 3×3 submatrices $\mathbf{P}_p$ and $\mathbf{P}_v$ are the error-covariance submatrices for the position and velocity components, respectively, and $\mathbf{P}_c$ is the error cross-covariance submatrix between position and velocity.

For 2D measurement data, $\mathbf{P}$ is configured as

$$\mathbf{P} = \begin{bmatrix} \sigma_{xx} & \sigma_{xy} & \sigma_{x\dot{x}} & \sigma_{x\dot{y}} \\ \sigma_{yx} & \sigma_{yy} & \sigma_{y\dot{x}} & \sigma_{y\dot{y}} \\ \sigma_{x\dot{x}} & \sigma_{x\dot{y}} & \sigma_{\dot{x}\dot{x}} & \sigma_{\dot{x}\dot{y}} \\ \sigma_{y\dot{x}} & \sigma_{y\dot{y}} & \sigma_{\dot{y}\dot{x}} & \sigma_{\dot{y}\dot{y}} \end{bmatrix} = \begin{bmatrix} \mathbf{P}_p & \mathbf{P}_c \\ \mathbf{P}_c^{\mathrm{T}} & \mathbf{P}_v \end{bmatrix}. \tag{5.48}$$

State estimate covariance can serve as a measure of how well a radar system meets its stated accuracy goal. For example, covariance analysis can be used to specify radar measurement accuracy over a number of measurements or time intervals via the **R** matrix, and to select the process noise covariance matrix **Q** that maintains sensitivity to maneuvers of the tracked object.

5.6 Relation of Measurement-to-Track Correlation Decision to the Kalman Gain

Since association is a statistical decision process, there will be detection and tracking errors due to closely spaced vehicles and clutter. The decision criterion, i.e., the gate, usually is constructed to yield a low probability of rejecting the correct measurement when it is present. Thus, a measurement $\mathbf{Z}_k$ is correlated with a track $\mathbf{X}_k$ if a number ξ_k can be found such that it is less than the gain G_k or the gate. When ξ_k is set equal to the normalized distance function, this statement is expressed mathematically as

$$\xi_k = [\mathbf{Z}_k - \mathbf{H}\hat{\mathbf{X}}_k]^{\mathrm{T}}[\mathbf{H}\hat{\mathbf{P}}_k\mathbf{H}^{\mathrm{T}} + \mathbf{R}_k]^{-1}[\mathbf{Z}_k - \mathbf{H}\hat{\mathbf{X}}_k] < G_k. \tag{5.49}$$

The notation $\hat{\mathbf{P}}_k$ is equivalent to $\mathbf{P}_{k|k-1}$ and $\hat{\mathbf{X}}_k$ is equivalent to $\hat{\mathbf{X}}_{k|k-1}$. The subscript k on **R** indicates that the measurement-noise covariance values may be a function of the sample number in general. The size of the gate G is found by requiring that

$$\text{Prob}\,[\text{Correct decision} \mid \mathbf{Z}_k \text{ present}] = \text{Prob}\,[\xi < G], \tag{5.50}$$

where the quadratic form of the test statistic ξ is distributed as a $\chi^2(n)$ random variable with the number of degrees of freedom n equal to the dimension of $\mathbf{Z}_k$. When $n = 2$,

$$\text{Prob}\,[\xi < G] = 1 - \exp(-G/2). \tag{5.51}$$

If $p_0 = \text{Prob}\,[\xi < G]$, then

$$G = -2\ln(1 - p_0). \tag{5.52}$$

A 2D system with $p_0 = 0.99$ yields a value of $G = 9.21$; for a 3D system with $p_0 = 0.99$, $G = 11.34$.[18]

Practically, measurements do not occur instantaneously because each measurement $\mathbf{Z}_i$ has an associated detection time t_i. Therefore, the gate test $\xi_{ij} < G$ for measurement $\mathbf{Z}_i$ against track $\mathbf{X}_j$ is defined by

$$\xi_{ij} = [\mathbf{Z}_i - \mathbf{H}\hat{\mathbf{X}}]^{\mathrm{T}}[\mathbf{H}\hat{\mathbf{P}}\mathbf{H}^{\mathrm{T}} + \mathbf{R}_i]^{-1}[\mathbf{Z}_i - \mathbf{H}\hat{\mathbf{X}}], \tag{5.53}$$

where

$$\hat{\mathbf{X}} = \hat{\mathbf{X}}_{t_i|t_{i-1}}, \tag{5.54}$$

$$\hat{\mathbf{P}} = \mathbf{F}(\Delta T)\hat{\mathbf{P}}_{t_i|t_{i-1}}[\mathbf{F}(\Delta T)]^{\mathrm{T}}, \tag{5.55}$$

and $\Delta T = t_i - t_{i-1}$.

5.7 Initialization and Subsequent Recursive Operation of the Kalman Filter

The following initialization process and equations are derived using the state transition and measurement models described earlier.[17] The Kalman filter is usually initialized with the first two measurements Z_0 and Z_1, where the measurements represent position. With this approach, the initial state estimate at the time of the second measurement Z_1 is

$$\mathbf{X}_1 = \begin{bmatrix} x_1 \\ v_1 \end{bmatrix} = \begin{bmatrix} Z_1 \\ \dfrac{Z_1 - Z_0}{\Delta T} \end{bmatrix}, \tag{5.56}$$

where ΔT is the time interval between measurements, x is the position of the object, and v is its speed.

The covariance of the state estimate $\mathbf{X}_1$ is given by

$$\mathbf{P}[\mathbf{X}_1] = \mathbf{P}_1 = \begin{bmatrix} \sigma_{xx} & \sigma_{xv} \\ \sigma_{xv} & \sigma_{vv} \end{bmatrix}_{k=1} = \begin{bmatrix} \sigma^2_{\varepsilon_1} & \sigma^2_{\varepsilon_1}(\Delta T)^{-1} \\ \sigma^2_{\varepsilon_1}(\Delta T)^{-1} & (\sigma^2_{\varepsilon_1} + \sigma^2_{\varepsilon_0})(\Delta T)^{-2} \end{bmatrix}, \tag{5.57}$$

where $\boldsymbol{\varepsilon}_k \sim \mathrm{N}[0, \mathbf{R}_k]$ is the measurement noise having a covariance matrix specified by the right-hand side of Eq. (5.57), whose terms correspond to Eqs. (5.77) through (5.79), and where $\mathbf{R}_k$ is of the form given by Eq. (5.24).[17] When the measurement noise is generated from a random sampling of the noise distribution $\boldsymbol{\varepsilon}_k$, the consistency of the filter initialization is guaranteed. If several Monte Carlo runs are made, random samples of the noise distribution $\boldsymbol{\varepsilon}_k \sim \mathrm{N}[0, \mathbf{R}_k]$ are taken for each run so that new and independent noises are incorporated into every run. Using the same initial conditions leads to biased estimates.[7]

The Kalman filter without process noise can be applied at this point to incorporate the subsequent measurements beginning with Z_2. The predicted state and predicted covariance at the time of the third measurement Z_2 are

$$\hat{\mathbf{X}}_2 = \mathbf{F}(\Delta T)\mathbf{X}_1 \tag{5.58}$$

and

$$\hat{\mathbf{P}}_2 = \mathbf{P}[\hat{\mathbf{X}}_2] = \mathbf{F}(\Delta T)\mathbf{P}_1[\mathbf{F}(\Delta T)]^{\mathrm{T}}, \tag{5.59}$$

where

$$\mathbf{F}(\Delta T) = \begin{bmatrix} 1 & \Delta T \\ 0 & 1 \end{bmatrix}, \tag{5.60}$$

as described in Section 5.11.1.

The Kalman gain applied to measurement Z_2 is

$$\mathbf{G}_2 = \begin{bmatrix} g_{x_2} \\ g_{v_2} \end{bmatrix} = \hat{\mathbf{P}}_2\mathbf{H}^{\mathrm{T}}[\mathbf{H}\hat{\mathbf{P}}_2\mathbf{H}^{\mathrm{T}} + \sigma_{\varepsilon_2}^2]^{-1} = \frac{1}{(\hat{\sigma}_{xx} + \sigma_{\varepsilon_2}^2)}\begin{bmatrix} \hat{\sigma}_{xx} \\ \hat{\sigma}_{xv} \end{bmatrix}, \tag{5.61}$$

where g_x and g_v are the gains pertaining to the position and velocity, respectively. The observation matrix $\mathbf{H}$ is

$$\mathbf{H} = [1 \;\; 0]. \tag{5.62}$$

It relates the state to the measurement according to Eq. (5.14) as

$$\mathbf{Z} = \mathbf{H}\begin{bmatrix} x \\ v \end{bmatrix} + \boldsymbol{\varepsilon}. \tag{5.63}$$

The covariance of the predicted state estimate and the Kalman gain are found by substituting the above expressions for $\mathbf{F}(\Delta T)$ and $\mathbf{P}_1$ into Eq. (5.59) and then $\hat{\mathbf{P}}_2$ and $\mathbf{H}$ into Eq. (5.61) to get

$$\hat{\mathbf{P}}_2 = \sigma_{\varepsilon_2}^2\begin{bmatrix} 5 & 3(\Delta T)^{-1} \\ 3(\Delta T)^{-1} & 2(\Delta T)^{-2} \end{bmatrix} \tag{5.64}$$

and

$$\mathbf{G}_2 = \begin{bmatrix} \frac{5}{6} & \frac{1}{2(\Delta T)} \end{bmatrix}^{\mathrm{T}}. \tag{5.65}$$

Finally, the corrected state and error-covariance values are given by

$$\mathbf{X}_2 = \hat{\mathbf{X}}_2 + \mathbf{G}_2(Z_2 - \mathbf{H}\hat{\mathbf{X}}_2) \tag{5.66}$$

and

$$\mathbf{P}_2 = (\mathbf{I} - \mathbf{G}_2\mathbf{H})\hat{\mathbf{P}}_2 = \sigma_{\varepsilon_2}^2 \begin{bmatrix} 5/6 & (2\Delta T)^{-1} \\ (2\Delta T)^{-1} & [2(\Delta T)^2]^{-1} \end{bmatrix}, \tag{5.67}$$

where $\mathbf{I}$ is the 2×2 identity matrix in this example, and

$$\mathbf{I} - \mathbf{G}_2\mathbf{H} = \begin{bmatrix} 1 - g_{x_2} & 0 \\ -g_{v_2} & 1 \end{bmatrix} = \begin{bmatrix} 1/6 & 0 \\ -(2\Delta T)^{-1} & 1 \end{bmatrix}. \tag{5.68}$$

After each prediction and correction update pair, the process repeats with the previous corrected estimates used to project or predict the new *a priori* estimates. Thus, Eqs. (5.58), (5.59), (5.61), (5.66), and (5.67) apply for any update k, where $k = 0, \ldots, M - 1$, and M is the number of measurements; i.e.,

$$\hat{\mathbf{X}}_{k+1} = \mathbf{F}(\Delta T)\mathbf{X}_k, \tag{5.69}$$

$$\hat{\mathbf{P}}_{k+1} = \mathbf{P}[\hat{\mathbf{X}}_{k+1}] = \mathbf{F}(\Delta T)\mathbf{P}_k[\mathbf{F}(\Delta T)]^{\mathrm{T}}, \tag{5.70}$$

$$\mathbf{G}_{k+1} = \begin{bmatrix} g_{x_{k+1}} \\ g_{v_{k+1}} \end{bmatrix} = \hat{\mathbf{P}}_{k+1}\mathbf{H}^{\mathrm{T}}[\mathbf{H}\hat{\mathbf{P}}_{k+1}\mathbf{H}^{\mathrm{T}} + \sigma_\varepsilon^2]^{-1} = \frac{1}{(\hat{\sigma}_{xx} + \sigma_\varepsilon^2)} \begin{bmatrix} \hat{\sigma}_{xx} \\ \hat{\sigma}_{xv} \end{bmatrix}, \tag{5.71}$$

$$\mathbf{X}_{k+1} = \hat{\mathbf{X}}_{k+1} + \mathbf{G}_{k+1}(\mathbf{Z}_{k+1} - \mathbf{H}\hat{\mathbf{X}}_{k+1}), \tag{5.72}$$

and

$$\mathbf{P}_{k+1} = (\mathbf{I} - \mathbf{G}_{k+1}\mathbf{H})\hat{\mathbf{P}}_{k+1}. \tag{5.73}$$

The notation $\hat{\mathbf{P}}_{k+1}$ is equivalent to $\mathbf{P}_{k+1|k}$, and $\mathbf{P}_{k+1}$ is equivalent to $\mathbf{P}_{k+1|k+1}$.

Recursive equations for the elements of the predicted error-covariance matrix $\hat{\mathbf{P}}_k$ at the time of any measurement k take the form

$$\hat{\sigma}_{xx} = \mathrm{E}[\hat{x}_k^2] = \frac{2(2k+1)}{k(k-1)}\sigma_\varepsilon^2, \tag{5.74}$$

$$\hat{\sigma}_{vx} = \mathrm{E}[\hat{x}_k\hat{v}_k] = \frac{6}{k(k-1)}\frac{\sigma_\varepsilon^2}{\Delta T}, \tag{5.75}$$

$$\hat{\sigma}_{vv} = \mathrm{E}[\hat{v}_k^2] = \frac{12}{k(k^2-1)}\frac{\sigma_\varepsilon^2}{(\Delta T)^2}, \tag{5.76}$$

while those for the corrected error-covariance matrix $\mathbf{P}_k$ are

$$\sigma_{xx} = \mathrm{E}[x_k^2] = \frac{2(2k-1)}{k(k+1)}\sigma_\varepsilon^2, \tag{5.77}$$

$$\sigma_{vx} = \mathrm{E}[x_k v_k] = \frac{6}{k(k+1)}\frac{\sigma_\varepsilon^2}{\Delta T}, \tag{5.78}$$

$$\sigma_{vv} = \mathrm{E}[v_k^2] = \frac{12}{k(k^2-1)}\frac{\sigma_\varepsilon^2}{(\Delta T)^2}. \tag{5.79}$$

The standard deviation of the measurements σ_ε is assumed to be constant in the above equations.

To obtain the estimated state at the time t_{k-1} of the last measurement, replace ΔT by $-\Delta T$. The Kalman gain equation in terms of g_x and g_v is useful for obtaining an initial estimate of tracking performance in terms of track time, i.e., the number of radar measurements or sample rate.[17]

Kalman filter gains may also be written as a function of the sample number k and the sampling interval ΔT by substituting Eqs. (5.74) through (5.76) into Eq. (5.71) as[17]

$$g_{x_k} = \frac{2(2k+1)}{(k+1)(k+2)} \tag{5.80}$$

and

$$g_{v_k} = \frac{6(\Delta T)^{-1}}{(k+1)(k+2)}. \tag{5.81}$$

When acceleration is present, the Kalman gain g_{a_k} for the acceleration component of the state is given by

$$g_{a_k} = \frac{12(\Delta T)^{-2}}{(k+1)^2(k+2)}. \tag{5.82}$$

Equations (5.80) through (5.82) show that the Kalman gains decrease asymptotically to zero as k becomes large. *This implies that the tracker, after a sufficient number of updates, will ignore the current and subsequent measurements and simply "dead reckon" the track based on past history.* The potential effects of this are reduced sensitivity to maneuvers such as a vehicle turning or changing lanes, creation of large lags or biases between the measurements and track position during and following a maneuver, and increased risk of track loss, particularly in clutter. Thus, gains should be large in order to weigh the current measurement more heavily than the past history when a change in direction of the tracked object is suspected. Therefore, typical implementations of the Kalman filter use **Q**-matrix process noise, precomputed gains related to the expected maneuver, or other methods to bound gains from below, and fixed gains after the desired track accuracy is achieved to prevent gains from growing without bound as discussed in Section 5.9.

In the majority of traffic management applications where vehicles appear on known roadway configurations, clutter may not pose as large an issue as in military applications where vehicles and other objects of interest are often found in forested or other high-clutter areas, or on unimproved roads without any foreknowledge of their whereabouts. Since roadway-mounted sensors operate at fixed locations and over relatively short ranges compared to their military counterparts, the detection area is smaller due to either smaller sensor footprints or smaller range-gated areas, making it easier to differentiate vehicles from roadside clutter. In addition, the locations of objects that contribute to roadside clutter are often detected during the calibration process for a traffic flow sensor and then subtracted from the scene during actual sensor operation.

5.8 The α-β Filter

5.8.1 Application and relation to Kalman gain

A widely used class of filters for finding the corrected state estimate has the form

$$\hat{\mathbf{X}}_{k+1|k+1} = \hat{\mathbf{X}}_{k+1|k} + \mathbf{G}_{k+1}(\mathbf{Z}_{k+1} - \hat{\mathbf{Z}}_{k+1|k}) \tag{5.83}$$

$$= \hat{\mathbf{X}}_{k+1|k} + \begin{bmatrix} \alpha \\ \dfrac{\beta}{\Delta T} \\ \dfrac{\gamma}{(\Delta T)^2} \end{bmatrix} (\mathbf{Z}_{k+1} - \hat{\mathbf{Z}}_{k+1|k}), \tag{5.84}$$

where the predicted measurement $\hat{\mathbf{Z}}_{k+1|k}$ can be rewritten as

$$\hat{\mathbf{Z}}_{k+1|k} = \mathbf{H}_{k+1}\hat{\mathbf{X}}_{k+1|k} \tag{5.85}$$

from the definitions in Eqs. (5.30) and (5.31).

When used to track an object moving with constant velocity, the filter is known as an α-β filter. When the acceleration component of the state is incorporated, the filter becomes an α-β-γ filter. Coefficients α, β, and γ are dimensionless, constant filter gains for the position, velocity, and acceleration, respectively. They are related to the Kalman gains presented in Section 5.7 by

$$g_x = \alpha, \tag{5.86a}$$

$$g_v = \frac{\beta}{\Delta T}, \tag{5.86b}$$

$$g_a = \frac{\gamma}{(\Delta T)^2}. \tag{5.86c}$$

Gains α, β, and γ are adjusted experimentally or, in the case of α and β, from relationships that express β in terms of α. By way of contrast, a Kalman filter's gain is computed recursively at each time step using the assumed process and measurement error statistics.

5.8.2 α-β filter equations for state estimate prediction and correction

Comparable to the Kalman filter equations, the α-β filter equations for an object moving with constant velocity consist of two parts: those that predict the state at the time of the next update and those that correct (update, filter) the state prediction.[19] The prediction equations are

$$x_p(k) = x_c(k-1) + v_c(k-1)\Delta T, \tag{5.87}$$

$$v_p(k) = v_c(k-1), \tag{5.88}$$

and the correction equations are

$$x_c(k) = x_p(k) + \alpha[x_m(k) - x_p(k)], \tag{5.89}$$

$$v_c(k) = v_p(k) + \frac{\beta}{\Delta T}[x_m(k) - x_p(k)], \tag{5.90}$$

where

$x_c(k)$ is the corrected position at the kth interval,
$x_p(k)$ is the predicted position at the kth interval,
$x_m(k)$ is the measured position at the kth interval,
$v_c(k)$ is the corrected velocity at the kth interval,
$v_p(k)$ is the predicted velocity at the kth interval,
ΔT is the radar update interval, and
α and β are filter coefficients.

5.8.3 Noise reduction and transient response properties of the α-β filter

Noise reduction is a property of the filter's response to a stochastic input from the measurement process.[19] The noise reduction ratio of the filter represents the mean-square error about the average measurement. The values of the position noise reduction ratio (NRR) and velocity noise reduction ratio are

$$\text{NRR}_x = \frac{2\alpha^2 - \beta(3\alpha - 2)}{\alpha(4 - 2\alpha - \beta)} \tag{5.91}$$

and

$$\text{NRR}_v = \frac{2\beta^2}{\Delta T\alpha(4 - 2\alpha - \beta)}, \tag{5.92}$$

respectively. For values of α and β used in tracking, these ratios are less than one, which reduces the uncertainty about the mean when using the corrected (filtered) measurements.

One can also determine the response of the filter to an unmodeled or a stochastic model of target behavior through the position transient response ratio (TRR) and the velocity transient response ratio, whose values are given by

$$\text{TRR}_x = \frac{(2 - \alpha)(1 - \alpha)^2\Delta T^2}{\alpha\beta(4 - 2\alpha - \beta)} \tag{5.93}$$

and

$$\text{TRR}_v = \frac{\alpha^2(2 - \alpha) + 2\beta(1 - \alpha)}{\alpha\beta(4 - 2\alpha - \beta)}, \tag{5.94}$$

respectively.

These equations demonstrate that the value of α has opposite results on transient response and on the effects of noise on sensor measurements. Small α produces more noise reduction in the measurements but increases the settling time from a transient, while large α produces less noise reduction and improves transient response. Therefore, larger α and β gains produce faster response for tracking transient changes, but provide smaller noise reduction in the measurements used to update the state. Assuming that a filter design should deal effectively with minimizing both the mean-square error associated with the sensor measurement and the filter response to unmodeled target behavior introduces a dilemma as these goals are contradictory.

5.8.4 Expressions for β as a function of α

A compromise between the contradictory goals is obtained through the Kalata relationship for β in terms of α, or $\beta = \beta(\alpha)$, which results in[20]

$$\beta = 2(2 - \alpha) - 4\sqrt{1 - \alpha}. \tag{5.95}$$

Gray[19] points out that Eq. (5.95) applies when uncertainty exists in both the position and velocity of the tracked object, but that the Benedict–Bordner relationship for $\beta = \beta(\alpha)$,

$$\beta = \frac{\alpha^2}{2-\alpha}, \tag{5.96}$$

applies when all of the uncertainty is in the position component of the state.

The limiting value of 0 for α and β implies that the measurement has no effect. For convergence and stability, the values of the α and β coefficients should be positive and small, for instance,

$$\left.\begin{array}{l} 0 < \alpha < 1 \\ 0 < \beta \le 2 \\ 0 < 4 - 2\alpha - \beta \end{array}\right\}. \tag{5.97}$$

5.9 Kalman Gain Control Methods

5.9.1 Preventing the gain from becoming too small

The **Q**-matrix method of preventing the gain from becoming too small injects a large value of process noise relative to the measurement noise covariance (the **R** matrix) into the state estimate prediction equation to drive the gains toward

$$\mathbf{G} = \begin{bmatrix} \mathbf{I} \\ \frac{1}{\Delta T}\mathbf{I} \end{bmatrix}, \tag{5.98}$$

where **I** is the 2 × 2 or 3 × 3 identity matrix, and ΔT is the time since the last update. Section 5.11 discusses several frequently utilized process noise models that assume the noise to be random and uncorrelated from sample to sample.

One method of adding noise through precomputed gains indexes the gains according to a noise-to-maneuver ratio (NMR) defined as

$$\mathrm{NMR} = \frac{2\sigma_\varepsilon}{a(\Delta T)^2}, \tag{5.99}$$

where a is the assumed acceleration, σ_ε is the standard deviation of the random measurement error, and ΔT is the update interval. Table 5.1 contains typical values of the position and velocity components of the Kalman gain as a function of NMR.[17]

The NMR approach may be applicable to specific types of vehicle tracking applications, e.g., tracking of drones and vehicles that change lanes or accelerate or decelerate when turning in a new direction. If we wish to add process noise in these applications with the NMR method, values of NMR range from 1.2 to 120 when $\sigma_\varepsilon = 0.3$ m [e.g., with location data from the Japanese Quasi-Zenith Satellite System (QZSS) designed to provide centimeter-level location accuracy], $a = 0.5$ m/s^2, and $0.1\ \mathrm{s} \le \Delta T \le 1\ \mathrm{s}$. If a is increased to 4 m/s^2, $0.15 \le \mathrm{NMR} \le 15$ when $\sigma_\varepsilon = 0.3$ m and $0.1\ \mathrm{s} \le \Delta T \le 1\ \mathrm{s}$. Values of

Table 5.1 Position and velocity components of Kalman gain versus NMR, where α and β are the components of the α-β filter described in Section 5.8.

NMR	$g_x = \alpha$ (position)	$g_v \times \Delta T = \beta$ (velocity)
$0 < \text{NMR} \leq 0.55$	1.0	1.0
$0.55 < \text{NMR} \leq 1.27$	0.9	0.6
$1.27 < \text{NMR} \leq 2.39$	0.8	0.5
$2.39 < \text{NMR} \leq 3.94$	0.71	0.43
$3.94 < \text{NMR} \leq 5.98$	0.64	0.21
$\text{NMR} > 5.98$	0.58	0.17

NMR range from 8 to 800 when $\sigma_\varepsilon = 2$ m [e.g., with a GPS receiver that incorporates the wide-area augmentation system (WAAS)], $a = 0.5$ m/s^2, and $0.1\text{ s} \leq \Delta T \leq 1\text{ s}$. If $a = 4$ m/s^2, $1 \leq \text{NMR} \leq 100$ when $\sigma_\varepsilon = 2$ m and $0.1\text{ s} \leq \Delta T \leq 1\text{ s}$. The values of a used in these calculations were estimated from data found in Hong, Lee, Borrelli, and Hedrick.[21] For these vehicle tracking applications, appropriate values of α and β may be inferred from Table 5.1 using the applicable ranges for NMR.

Other methods proposed by Hu, Chen, Chen, and Liu[22] to ensure that data from sensor measurements, such as from a GPS device, are not ignored to the detriment of the state estimation process are to either (1) apply a factor $S > 1$ to the predicted covariance matrix given by Eq. (5.42) or (5.46) to increase the variance of the predicted state vector and thus give more weight to the measurements or (2) estimate the variance of the state transition model directly from the predicted innovation.

The first technique takes the form of a fading-memory algorithm, where the predicted covariance matrix is written as

$$\mathbf{P}_{k+1|k} = S(\mathbf{F}\mathbf{P}_{k|k}\mathbf{F}^{\mathrm{T}} + \mathbf{Q}). \tag{5.100}$$

The scale factor S may be selected as a constant between 1 and ≈1.4. When $S = 1$, the technique reverts to the conventional Kalman filter. There are some drawbacks to the constant scale factor scheme. For example, as the filtering proceeds as additional measurements become available, it increasingly ignores old data. Therefore, Hu et al.[22] propose a variable scale factor based on the accuracy of the state transition and measurement models. One configuration for their variable scale factor S is given by

$$S \geq \frac{\dfrac{1}{m_{k+1}}\left(\mathbf{w}_{k+1|k}^{\mathrm{T}}\mathbf{w}_{k+1|k}\right)}{\dfrac{1}{N}\displaystyle\sum_{j=0}^{N-1}\dfrac{1}{m_{k-N+j}}\left(\mathbf{w}_{k-N+j|k-N}^{\mathrm{T}}\mathbf{w}_{k-N+j|k-N}\right)}, \tag{5.101}$$

where $\mathbf{w}_k$ is the innovation, N is the number of previous samples (also called the filtering window), j is an index between 0 and $N - 1$, and m_i is the number of measurements at epoch i. A value of $S \leq 1$ indicates that the filtering process is in steady state. When $S > 1$, the filtering may be in an unstable state or in a state of divergence. In practice then, a threshold $S_0 > 1$ is selected in order to avoid false alarms. When $S > S_0$, the adaptive algorithm is applied. Otherwise, $S = 1$, and the conventional Kalman filtering algorithm is used.

The second approach estimates only the variance of the state transition model since GPS measurement noise can be determined to a reasonable level knowing the type of GPS receiver used. Hu et al.[22] express the Kalman filter equation as a Gauss–Markov model and proceed to determine the covariance matrix of the predicted residuals.

As might be expected, the dynamic noise level strongly affects the effectiveness of the conventional Kalman filter, but with the fading-memory filter, the positioning errors are significantly reduced. However, the positioning accuracy is still affected by the selection of initial dynamic noise level. The variance estimation filter exhibits better performance and is not affected by the selection of the initial dynamic noise level.

5.9.2 Preventing the gain from becoming too large

A limit must be imposed on the amount of added process noise because an unbounded increase of **Q**-matrix noise almost surely results in a clutter-to-measurement correlation. Moreover, large values of **Q**-matrix parameters cause large gains, i.e., near unity for the position submatrix. The large gains shift the position variables in the state estimate to the measurement values, and a radical change in the velocity vector occurs. This can lead to tracking of clutter measurement data and ultimately result in track loss on the *bona fide* object of interest.[17]

One method of limiting the gain is by using the trace of the $\mathbf{P}_v$ submatrix in Eq. (5.47) or (5.48) as a measure of the track accuracy. Accordingly,

$$\text{if Trace}(\mathbf{P}_v) < \text{Goal, then set } g_v = 2g_v \text{ or equivalently, } \beta = 2\beta \quad (5.102)$$

in the equation for the Kalman gain. This will cause $\mathbf{P}_{v_{k|k}}$ to remain constant on subsequent updates and for $\mathbf{P}_{p_{k|k}}$ to decrease slightly for the next two or three updates. An alternative measure for triggering the increase of g_v to $2g_v$ is to compare the maximum values of the position error covariances with the goal value of the noise as follows:

$$\text{if } \max[\sigma_{xx}\sigma_{yy}\sigma_{zz}] < \text{Goal, then set } g_v = 2g_v, \text{ or } \beta = 2\beta. \quad (5.103)$$

These techniques will clamp the gains within two to three updates after the goal is achieved since the corrected covariance and, therefore, the prediction covariance, are approximately constant.

5.10 Noise Covariance Values and Filter Tuning

In the actual implementation of the Kalman filter, it is usually possible to obtain the measurement noise covariance values that appear in **R** prior to operation of the filter since the process must be observed and measured anyway while operating the filter. Therefore, it should be practical to perform offline measurements in order to determine the variance of the measurement noise if it is not already provided by the manufacturer of the radar or another measurement system.

Determining the process noise covariance values in **Q** is normally more difficult because, typically, it is not possible to directly observe the process being estimated. Sometimes a relatively simple (i.e., poor or unsophisticated) process model can produce acceptable results if one injects enough uncertainty into the process via the **Q** matrix as described above and in the next sections.

In either case, whether or not there is a rational basis for choosing the parameters, superior filter performance (statistically speaking) can often be obtained by tuning the filter parameters in the **Q** and **R** matrices. The tuning is usually performed offline, frequently with the help of another (distinct) Kalman filter in a process referred to as system identification.

Under conditions where the **Q** and **R** matrices are constant, both the estimation error covariance and the Kalman gain will stabilize quickly and then remain constant. If this is the case, these parameters can be precomputed by either running the filter offline or, for example, by determining the steady state value of $\mathbf{P}_{k+1}$ as described above and by Grewal and Andrews.[6]

In other applications, however, the measurement error in particular does not remain constant. For example, when sighting beacons in optoelectronic tracker ceiling panels, the noise in measurements of nearby beacons will be smaller than that in distant beacons. Also, the process noise is sometimes changed dynamically during filter operation—becoming a function of the sampling time as $\mathbf{Q}_k$—in order to adjust to different dynamics. A nonradar example of this effect occurs when tracking the head of a user of a 3D virtual environment. Here the magnitude of $\mathbf{Q}_k$ may be reduced if the user appears to be moving slowly, but may be increased if the dynamics start changing rapidly. In such cases, $\mathbf{Q}_k$ might be chosen to account for both uncertainty about the user's intentions and uncertainty in the model.[5]

5.11 Process Noise Covariance Matrix Models

For many applications of Kalman filtering, a good rationale does not exist for selecting the values in the process noise covariance matrix **Q**. In those cases, rules of thumb are resorted to that include the use of simple models, empirical data, or anything else that appears to give a satisfactory solution. For many vehicle-tracking and maneuver prediction[23,24] applications, the constant-velocity and constant-acceleration object process noise models discussed below appear to be applicable. The common assumption in these models is that the noise is random and uncorrelated from sample to sample. Other models, such as the Singer model, assume correlated noise from sample to sample as described in Blackman[3] and Singer.[12] Still other process noise models are summarized by Li and Jilkov.[13]

5.11.1 Constant-velocity object process noise model

If the object of interest is moving at constant velocity, i.e., the acceleration is nominally zero, then a generic coordinate x may be described by[7]

$$\ddot{x}(t) = 0. \tag{5.104}$$

In the absence of noise, the position $x(t)$ evolves according to a polynomial in time. In practice, the velocity undergoes small changes due to continuous-time white noise w, resulting in an acceleration denoted as

$$\ddot{x}(t) = w(t), \tag{5.105}$$

where

$$\mathrm{E}[w(t)] = 0, \tag{5.106}$$

$$\mathrm{E}[w(t)w(\tau)] = q(t)\delta(t - \tau), \tag{5.107}$$

q is the variance of $w(t)$, and δ is the Dirac delta function.

The state vector corresponding to Eq. (5.105) is

$$\mathbf{X} = \begin{bmatrix} x \\ \dot{x} \end{bmatrix}. \tag{5.108}$$

In many applications, the model of Eq. (5.104) is utilized for each coordinate. Furthermore, the motion along each coordinate is assumed to be decoupled from the others, and the noises entering each component are assumed to be mutually independent with potentially different and time-varying intensities.

The continuous-time state equation is

$$\dot{x}(t) = \mathbf{AX}(t) + \begin{bmatrix} 0 \\ 1 \end{bmatrix} w(t), \tag{5.109}$$

where

$$\mathbf{A} = \begin{bmatrix} 0 & 1 \\ 0 & 0 \end{bmatrix}. \tag{5.110}$$

Since the following relations apply to linear time-invariant systems,[7,15]

$$\mathbf{F}_k \equiv \mathbf{F}(t_{k+1},\ t_k) = \mathbf{F}(t_{k+1} - t_k) = \mathrm{e}^{\mathrm{A}(t_{k+1}-t_k)}, \tag{5.111}$$

$$\mathbf{w}_k \equiv \mathbf{w}(t_k) = \int_{t_k}^{t_{k+1}} \mathrm{e}^{\mathrm{A}(t_{k+1}-\tau)}\mathbf{w}(\tau)\mathrm{d}\tau, \tag{5.112}$$

and

$$\mathbf{Q}_k = \int_{t_k}^{t_{k+1}} \mathrm{e}^{\mathrm{A}(t_{k+1}-\tau)}\mathbf{w}(\tau)\mathrm{e}^{\mathrm{A}^{\mathrm{T}}(t_{k+1}-\tau)}\mathrm{d}\tau, \tag{5.113}$$

we can write the state equation of Eq. (5.109) in terms of the sample number k, as was done in Eq. (5.3) (rather than as a continuous-time equation having a time-based index) as

$$\mathbf{X}_{k+1} = \mathbf{F}\mathbf{X}_k + \mathbf{w}_k, \tag{5.114}$$

where the state transition matrix is

$$\mathbf{F} = \mathrm{e}^{\mathbf{A}\Delta T} = \begin{bmatrix} 1 & \Delta T \\ 0 & 1 \end{bmatrix}, \tag{5.115}$$

and the discrete-time process noise is related to the continuous-time process noise through

$$\mathbf{w}_k = \int_0^{\Delta T} \mathrm{e}^{\mathbf{A}(\Delta T-\tau)} \begin{bmatrix} 0 \\ 1 \end{bmatrix} \mathbf{w}[k(\Delta T) - \tau]\mathrm{d}\tau. \tag{5.116}$$

The derivation of the state transition matrix appearing in Eq. (5.115) is found in the Appendix to this book.

Utilizing Eq. (5.107) and assuming that q is constant allows the covariance matrix for $\mathbf{w}_k$ to be expressed as

$$\mathbf{Q} = \mathrm{E}[\mathbf{w}_k\mathbf{w}_k^{\mathrm{T}}] = q\int_0^{\Delta T} \begin{bmatrix} \Delta T - \tau \\ 1 \end{bmatrix} [\,\Delta T - \tau \quad 1\,]\mathrm{d}\tau = q\begin{bmatrix} \frac{1}{3}(\Delta T)^3 & \frac{1}{2}(\Delta T)^2 \\ \frac{1}{2}(\Delta T)^2 & \Delta T \end{bmatrix}. \tag{5.117}$$

Scale factor q is defined such that the change in velocity over the sampling interval ΔT is on the order of

$$\sqrt{Q_{22}} = \sqrt{q(\Delta T)}. \tag{5.118}$$

5.11.2 Constant-acceleration object process noise model

Derivation of the **Q** matrix for an object moving with constant acceleration closely follows the derivation for the constant-velocity object. The constant-acceleration object for a generic coordinate x is described by

$$\dddot{x}(t) = 0. \tag{5.119}$$

As in Eq. (5.105), the acceleration is never exactly constant, and its slight changes are modeled by zero-mean white noise as

$$\dddot{x}(t) = w(t). \tag{5.120}$$

The smaller the variance q of $w(t)$, the more nearly constant is the acceleration. The state vector corresponding to Eq. (5.120) is

$$\mathbf{X} = \begin{bmatrix} x \\ \dot{x} \\ \ddot{x} \end{bmatrix}, \tag{5.121}$$

and its continuous-time state equation is

$$\dot{x}(t) = \mathbf{AX}(t) + \begin{bmatrix} 0 \\ 0 \\ 1 \end{bmatrix} w(t), \tag{5.122}$$

where

$$\mathbf{A} = \begin{bmatrix} 0 & 1 & 0 \\ 0 & 0 & 1 \\ 0 & 0 & 0 \end{bmatrix}. \tag{5.123}$$

The discrete-time state equation with sampling interval ΔT is identical to that in Eq. (5.114) but with

$$\mathbf{F} = \mathrm{e}^{\mathbf{A}\Delta T} = \begin{bmatrix} 1 & \Delta T & \frac{1}{2}(\Delta T)^2 \\ 0 & 1 & \Delta T \\ 0 & 0 & 1 \end{bmatrix} \tag{5.124}$$

and the covariance matrix of $\mathbf{w}_k$ as

$$\mathbf{Q} = \mathrm{E}[\mathbf{w}_k \mathbf{w}_k^{\mathrm{T}}] = q \begin{bmatrix} \frac{1}{20}(\Delta T)^5 & \frac{1}{8}(\Delta T)^4 & \frac{1}{6}(\Delta T)^3 \\ \frac{1}{8}(\Delta T)^4 & \frac{1}{3}(\Delta T)^3 & \frac{1}{2}(\Delta T)^2 \\ \frac{1}{6}(\Delta T)^3 & \frac{1}{2}(\Delta T)^2 & \Delta T \end{bmatrix}. \tag{5.125}$$

Scale factor q is defined in this case such that the change in acceleration over the sampling interval ΔT is on the order of

$$\sqrt{Q_{33}} = \sqrt{q(\Delta T)}. \tag{5.126}$$

5.12 Interacting Multiple Model for Vehicle Motion on a Roadway

Barrios and Motai propose an interacting multiple model (IMM) to represent common vehicle movement scenarios.[24] The four models in their system are constant (vehicle) location (CL), constant velocity (CV), constant acceleration (CA), and constant jerk (CJ), representing vehicle motion with a constant change in acceleration. Each of these kinematic models is implemented as a Kalman filter in an IMM that estimates a vehicle's trajectory 3 s into the future. The 3-s window doubles the average human reaction time of 1.5 s to stop a vehicle and thus allows for slower reaction times as well as variations caused by the speed and type of vehicle because a faster or heavier vehicle will need additional time to slow down and stop.

The IMM state estimate is improved by using GPS location data to correct a poor estimate, particularly on curves. Simulations of the IMM show that when a vehicle enters a turn, the prediction of its future location is erroneous; i.e., it often places the vehicle outside of the road boundaries. If the system knew the direction of the road ahead and whether the estimated future location of the vehicle was within the road boundaries, the accuracy and reliability of the state estimation process would be improved. When GPS data were made available, the vehicle location estimate did improve, providing that the vehicle remained on the road, the driver was operating the vehicle properly, and the driver was alert and able to follow GPS guidance.

5.12.1 Kinematic models

The CL model is expressed as

$$x_k = x_{k-1} + w\Delta T, \tag{5.127}$$

$$v_k = 0, \tag{5.128}$$

where x represents position, and v represents velocity. The CV model is expressed as

$$x_k = x_{k-1} + [v_{k-1} + w]\Delta T, \tag{5.129}$$

$$v_k(k) = v_{k-1} + w. \tag{5.130}$$

The CA model is expressed as

$$x_k = x_{k-1} + v_{k-1}\Delta T + \frac{1}{2}[a_{k-1} + w](\Delta T)^2, \tag{5.131}$$

$$v_k = v_{k-1} + [a_{k-1} + w]\Delta T. \tag{5.132}$$

The CJ model is expressed as

$$x_k = x_{k-1} + v_{k-1}\Delta T + \frac{1}{2}a_{k-1}(\Delta T)^2 + \frac{1}{6}[j_{k-1} + w](\Delta T)^3, \tag{5.133}$$

$$v_k = v_{k-1} + a_{k-1}\Delta T + \frac{1}{2}[j_{k-1} + w](\Delta T)^2. \tag{5.134}$$

5.12.2 IMM implementation

In the Barrios and Motai example, the IMM system of Fig. 5.4 contained the four models defined above. The progression through the IMM algorithm is described by the following four steps:[25,26]

1. Matched filters for each model are run in parallel, yielding the state estimate conditioned on each model being the current one.
2. The current probability of each model is evaluated in a Bayesian framework using the likelihood function of each filter.
3. Each filter's input at the beginning of the cycle is a combination of its output from the previous cycle with suitable weightings that reflect both the probability of each model being the current one and the model transition probabilities.
4. The combined state estimate and error-covariance outputs are computed using the current model probabilities.

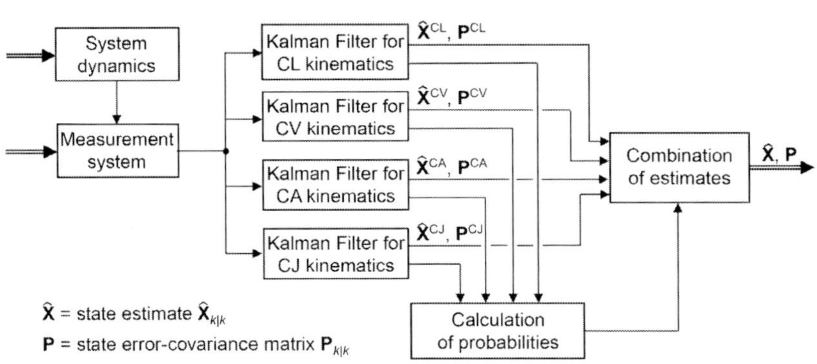

Figure 5.4 Interacting multiple-model algorithm with constant location (CL), constant velocity (CV), constant acceleration (CA), and constant jerk (CJ) kinematic models.

A model M consists of a state transition matrix $\mathbf{F}$ and a representation for the process noise covariance values $\mathbf{Q}$ as defined in Eqs. (5.4) through (5.6). The likelihood function for a model that is in effect at a particular sampling time, the output state estimate, and the error covariance are calculated as follows.[7,25,26]

Let M^j be the event that model j is correct with prior probability

$$P[M^j] = \mu^j_{k=0},\ j = 1,\ \ldots,\ r, \tag{5.135}$$

where k is the sample number, and r is the number of models.

The likelihood function of the measurements up to sample k under the assumption that model j is activated is

$$\lambda^j_k = P(\mathbf{Z}_k|M^j) = \prod_{i=1}^{k} p[w^j_i], \tag{5.136}$$

where the probability density function of the innovation from filter j, assuming a Gaussian distribution, is

$$p[\mathbf{w}^j_k] = |2\pi\mathbf{S}^j_k|^{-1/2}\exp\left[-\frac{1}{2}(\mathbf{w}^j_k)^{\mathrm{T}}(\mathbf{S}^j_k)^{-1}\mathbf{w}^j_k\right], \tag{5.137}$$

and where $\mathbf{w}_k$ is defined in Eq. (5.30).

Applying Bayes' theorem gives the posterior probability that model j is correct at time k when measurement $\mathbf{Z}_k$ occurs as

$$\mu^j_k \equiv P\,[M^j|\mathbf{Z}_k] = \frac{p[w^j_k]\mu^j_{k-1}}{\sum_{l=1}^{r} p\,[w^l_k]\mu^l_{k-1}} = \frac{\lambda^j_k\,\mu^j_{k=0}}{\sum_{l=1}^{r}\lambda^l_k\,\mu^l_{k=0}}. \tag{5.138}$$

The above derivation is exact under the following two assumptions:

1. The correct model is among the set of r models under consideration.
2. The same model has been in effect from the initial time.

The first assumption is a reasonable approximation. However, the second is not if the maneuver has started at some time within the interval $[1, k]$. Hence, a heuristic approach of creating a lower bound for each model's probability may be adopted, enabling this technique to track switching models. Alternatively, a sliding-window or fading-memory likelihood function of the form

$$\lambda^j_k = [\lambda^j_{k-1}]^{\kappa} p\,(w^j_k)\ \text{ for } 0 < \kappa < 1 \tag{5.139}$$

can be used.[7]

The output state estimate is a weighted average of the model-conditioned estimates with the probabilities of Eq. (5.138) used as weights. Thus,

$$\hat{\mathbf{X}}_{k|k} = \mathrm{E}[\mathbf{X}_k|\mathbf{Z}_k] = \sum_{j=1}^{r} \mathrm{E}[\mathbf{X}_k|M^j, \mathbf{Z}_k]\, P\,[M^j|\mathbf{Z}_k] \tag{5.140}$$

$$\begin{aligned} &= \sum_{j=1}^{r} \mathrm{E}[\mathbf{X}_k|M^j, \mathbf{Z}_k, \mathbf{X}_{k-1|k-1}, \mathbf{P}_{k-1|k-1}]\, \mu_k^j \\ &= \sum_{j=1}^{r} \mathbf{X}_{k|k}^j\, \mu_k^j, \end{aligned} \tag{5.141}$$

where $P[M^j \mid \mathbf{Z}^k]$ is the probability that model j is in effect at time $k \bullet \Delta T$ when measurement $\mathbf{Z}_k$ is made, and $\mathbf{X}_{k-1|k-1}$, $\mathbf{P}_{k-1|k-1}$ are known prior to k.

The output state error covariance is expressed as

$$\mathbf{P}_{k|k} = \sum_{j=1}^{r} \mu_k^j \left\{ \mathbf{P}_{k|k}^j + [\hat{\mathbf{X}}_{k|k}^j - \hat{\mathbf{X}}_{k|k}][\hat{\mathbf{X}}_{k|k}^j - \hat{\mathbf{X}}_{k|k}]^{\mathrm{T}} \right\}, \tag{5.142}$$

where the measurement innovation (residual) is given by Eq. (5.31); $\mathbf{S}_k^j$ is given by Eq. (5.32); $\mathbf{P}_{k|k}^j$ is given by Eq. (5.33), (5.34), or (5.35); $\mathbf{G}_k^j$ is given by Eq. (5.36); $\mathbf{X}_{k|k}^j$ is given by Eq. (5.37) or (5.38); and $\mathbf{P}_{k|k}^j$ is given by Eq. (5.39).

The combined estimates in Eq. (5.140) or (5.141) and Eq. (5.142) are the minimum mean-square error (MMSE) estimates computed probabilistically over all of the models. By assumption, one of the models is the correct model. Therefore, one may simply use the estimate from the model with the largest value of posterior probability μ_k^j to eliminate those models with low probabilities, or adopt some other *ad hoc* method of model selection.

5.12.3 Test results

Barrios and Motai[24] calculated the average speed from the location difference provided by the current and previous GPS location measurements and the GPS update period ΔT of 1 s. The acceleration was calculated from the velocity difference between the current and previous GPS measurements and the jerk value from the acceleration difference between the current and previous values, along with the update period of 1 s.

The measurement-noise covariance matrix $\mathbf{R}$ was found from the covariance of the differences in the measurements, while the process-noise covariance matrix $\mathbf{Q}$ for each Kalman filter model was obtained from the covariance of the differences in the measurements compared with their respective x and y components. Using these results, Barrios and Motai[24] were able to compute the transition probability matrix for the IMM as

$$\mu_k^j = \begin{bmatrix} CL \to CL & CL \to CV & CL \to CA & CL \to CJ \\ CV \to CL & CV \to CV & CV \to CA & CV \to CJ \\ CA \to CL & CA \to CV & CA \to CA & CA \to CJ \\ CJ \to CL & CJ \to CV & CJ \to CA & CJ \to CJ \end{bmatrix} \tag{5.143}$$

$$= \begin{bmatrix} 0.154 & 0.154 & 0.385 & 0.308 \\ 0.011 & 0.470 & 0.305 & 0.214 \\ 0.014 & 0.259 & 0.458 & 0.269 \\ 0.002 & 0.243 & 0.508 & 0.247 \end{bmatrix}. \tag{5.144}$$

The road configuration selected to evaluate the combined IMM-GPS state estimation method had a sharp turn that was negotiated at a speed of ~60 km/h. The GPS receiver was equipped with a WAAS and had an accuracy better than 2.2 m horizontal 95% of the time. The results show that the average errors for the predicted vehicle's location 1 s and 3 s ahead were 1.9 m and 5.1 m, respectively, over the total test route. Improvements to future applications of this technique include use of a GPS with a higher update frequency and smaller location error, which will be available as new GPS frequencies come online.

5.13 Extended Kalman Filter

The EKF is often used when nonlinearities are present in the state transition process to be estimated and updated, in the observation matrix, or in the covariance matrices of the noise sources. Nonlinear motion of objects is common in radar tracking of slowly turning objects with high–data-rate radars (e.g., more than ten updates during a maneuver). Practically, almost no tracking problem is truly linear. Furthermore, additional nonlinearities arise when the measurement-space and tracking-space coordinate systems are different.

The EKF linearizes the state estimation process about the current mean and covariance of the state using first-order Taylor approximations to the time-varying transition and observation matrices, assuming that the parameters of the nonlinear dynamical system, namely, $\mathbf{F}(t)$, $\mathbf{H}(t)$, $\mathbf{Q}(t)$, $\mathbf{R}(t)$, μ_1, and σ_1, are known. The parameters μ_1 and σ_1 are the mean and variance, respectively, of the normally distributed initial state $\mathbf{X}_1$. *Unlike its linear counterpart, the EKF is not an optimal estimator of nonlinear processes.* In addition, if the initial estimate of the state is wrong or if the process is modeled incorrectly, the filter may quickly diverge because of its method of linearization. Another issue with the EKF is that the estimated state error-covariance matrix tends to underestimate the true covariance matrix and therefore risks becoming inconsistent in the statistical sense without the addition of stabilizing noise as described in Sections 5.9 and 5.11.

To derive the extended Kalman filter, we can express the nonlinear state transition equation as[7,14,27]

$$\mathbf{X}_{k+1} = f_k \mathbf{X}_k + \mathbf{w}_k, \tag{5.145}$$

where $f:\Re^{n_x} \rightarrow \Re^{n_x}$ is a nonlinear function that replaces the state transition matrix $\mathbf{F}$ found in the equations for the standard Kalman filter. Without loss of generality, this derivation assumes that the system has no external input (control function).[14] The nonlinear equation relating the state to measurements is given by

$$\mathbf{Z}_{k+1} = h(\mathbf{X}_{k+1}) + \boldsymbol{\varepsilon}_{k+1}, \tag{5.146}$$

where $h:\Re^{n_z} \rightarrow \Re^{n_z}$ is a nonlinear function that replaces the observation matrix $\mathbf{H}$.

If a state estimate is available at time t_k, then the estimated state $\hat{\mathbf{X}}_k$, given measurements up to and including time t_k, may be written as

$$\hat{\mathbf{X}}_k = \mathrm{E}[\mathbf{X}_k | z_1,\ldots, z_k], \tag{5.147}$$

where $\mathbf{X}_k$ is the actual state,

$$\mathrm{E}[\hat{\mathbf{X}}_k] = \mathbf{X}_k, \text{ and} \tag{5.148}$$

$$\mathrm{cov}[\hat{\mathbf{X}}_k] = \mathrm{E}[(\hat{\mathbf{X}}_k - \mathbf{X}_k)(\hat{\mathbf{X}}_k - \mathbf{X}_k)^{\mathrm{T}}] = \mathbf{P}_k. \tag{5.149}$$

Process and measurement noise have the statistics found in Eqs. (5.5), (5.6), (5.12), and (5.13) as before.

Using Taylor's theorem to linearize the system dynamics $\mathbf{X}_{k+1} = f_k \mathbf{X}_k + \mathbf{w}_k$ around $\hat{\mathbf{X}}_{k|k}$ leads to a state transition equation of the form

$$\hat{\mathbf{X}}_{k+1|k} = f(\hat{\mathbf{X}}_{k|k}) + [\nabla_x f(\hat{\mathbf{X}}_{k|k})](\mathbf{X}_k - \hat{\mathbf{X}}_{k|k}) + \text{higher-order terms}, \tag{5.150}$$

where

$$\nabla_x f(\mathbf{X}) = \frac{\partial f_k(x)}{\partial x_k}, \tag{5.151}$$

which is a Jacobian matrix of partial derivatives of $f(\cdot)$ with respect to x,

$$\hat{\mathbf{X}}_{k+1|k} = \mathrm{E}[\mathbf{X}_{k+1|k} | z_1,\ldots, z_k] = f(\hat{\mathbf{X}}_{k|k}), \tag{5.152}$$

$$
\begin{aligned}
\hat{\mathbf{P}}_{k+1|\mathrm{k}} &= \mathrm{cov}[\hat{\mathbf{X}}_{k+1|k}] = \mathrm{E}[(\hat{\mathbf{X}}_{k+1|k} - \mathbf{X}_{k+1})(\hat{\mathbf{X}}_{k+1|k} - \mathbf{X}_{k+1})^{\mathrm{T}}] \\
&= [\nabla_x f(\hat{\mathbf{X}}_{k|k})]\mathrm{E}[(\hat{\mathbf{X}}_{k|k} - \mathbf{X}_k)(\hat{\mathbf{X}}_{k|k} - \mathbf{X}_k)^{\mathrm{T}}][\nabla_x f(\hat{\mathbf{X}}_{k|k})]^{\mathrm{T}} \\
&= [\nabla_x f(\hat{\mathbf{X}}_{k|k})]\hat{\mathbf{P}}_{k|k}[\nabla_x f(\hat{\mathbf{X}}_{k|k})]^{\mathrm{T}} = \mathbf{F}_x\hat{\mathbf{P}}_{k|k}\mathbf{F}_x^{\mathrm{T}},
\end{aligned}
\tag{5.153}
$$

$$
\mathbf{F}_x = [\nabla_x f(\mathbf{X})] = \begin{bmatrix} \dfrac{\partial f_1(x)}{\partial x_1} & \dots & \dfrac{\partial f_1(x)}{\partial x_n} \\ \vdots & \dots & \vdots \\ \dfrac{\partial f_n(x)}{\partial x_1} & \dots & \dfrac{\partial f_n(x)}{\partial x_n} \end{bmatrix}, \tag{5.154}
$$

$$
\mathbf{X}^{\mathrm{T}} = [x_1 \cdots x_n], \text{ and} \tag{5.155}
$$

$$
f^{\mathrm{T}} = [f_1 \cdots f_n]. \tag{5.156}
$$

Next we compute the predicted state from the actual nonlinear function that appears in Eq. (5.152), and the predicted state error-covariance matrix and the matrix of partial derivatives $\mathbf{F}_x$ from Eqs. (5.153) and (5.154) to obtain

$$
\mathbf{P}_{k+1|k} = \mathbf{F}_x\mathbf{P}_{k|k}\mathbf{F}_x^{\mathrm{T}} + \mathbf{Q}_k. \tag{5.157}
$$

Similarly, we apply Taylor's theorem to linearize the observation function h that relates the measurements to the state through

$$
\mathbf{Z}_{k+1} = h(\mathbf{X}_{k+1}) + \boldsymbol{\varepsilon}_{k+1} \tag{5.158}
$$

to find that

$$
\mathbf{Z}_{k+1} = h(\hat{\mathbf{X}}_{k+1}) + [\nabla_x h(\hat{\mathbf{X}}_{k+1})](\mathbf{X}_{k+1} - \hat{\mathbf{X}}_{k+1}) + \text{higher-order terms}, \tag{5.159}
$$

where

$$
\mathbf{Z}_{k+1} = h(\hat{\mathbf{X}}_{k+1}) + \mathbf{H}_x[\mathbf{X}_{k+1} - \hat{\mathbf{X}}_{k+1}], \text{ and} \tag{5.160}
$$

$$
\mathbf{H}_x = [\nabla_x h(x)] = \begin{bmatrix} \dfrac{\partial h_1(x)}{\partial x_1} & \dots & \dfrac{\partial h_1(x)}{\partial x_n} \\ \vdots & \dots & \vdots \\ \dfrac{\partial h_m(x)}{\partial x_1} & \dots & \dfrac{\partial h_m(x)}{\partial x_n} \end{bmatrix}. \tag{5.161}
$$

Then we calculate the predicted measurement from the actual nonlinear function as

$$\hat{\mathbf{Z}}_{k+1} = h(\hat{\mathbf{X}}_{k+1}). \tag{5.162}$$

Finally, we determine the predicted gain using the matrix of partial derivatives $\mathbf{H}_x$ and the corrected (updated or filtered) state and error-covariance matrices as

$$\mathbf{G}_{k+1} = \mathbf{P}_{k+1|k}\mathbf{H}_x^{\mathrm{T}}[\mathbf{H}_x\mathbf{P}_{k+1|k}\mathbf{H}_x^{\mathrm{T}} + \mathbf{R}_k]^{-1}, \tag{5.163}$$

$$\hat{\mathbf{X}}_{k+1|k+1} = \hat{\mathbf{X}}_{k+1|k} + \mathbf{G}_{k+1}(\mathbf{Z}_{k+1} - \mathbf{H}_x\hat{\mathbf{X}}_{k+1|k}), \tag{5.164}$$

$$\begin{aligned}\mathbf{P}_{k+1|k+1} &= [\mathbf{I} - \mathbf{G}_{k+1}\mathbf{H}_x]\mathbf{P}_{k+1|k}[\mathbf{I} - \mathbf{G}_{k+1}\mathbf{H}_x]^{\mathrm{T}} + \mathbf{G}_{k+1}\mathbf{R}_k\mathbf{G}_{k+1}^{\mathrm{T}} \\ &= [\mathbf{I} - \mathbf{G}_{k+1}\mathbf{H}_x]\mathbf{P}_{k+1|k}.\end{aligned} \tag{5.165}$$

In summary, an iteration of the EKF for nonlinear state transition and observation functions contains the following steps:

1. Begin with the last corrected state estimate $\hat{\mathbf{X}}_{k|k}$.
2. Linearize the system dynamics $\mathbf{X}_{k+1} = f_k\mathbf{X}_k + \mathbf{w}_k$ around $\hat{\mathbf{X}}_{k|k}$.
3. Apply the prediction step of the Kalman filter to the linearized system dynamic equation of Step 2 to get $\hat{\mathbf{X}}_{k+1|k}$ and $\mathbf{P}_{k+1|k}$.
4. Linearize the observation dynamics $\mathbf{Z}_{k+1} = h(\mathbf{X}_{k+1}) + \boldsymbol{\varepsilon}_{k+1}$ around $\hat{\mathbf{X}}_{k+1|k}$.
5. Apply the correction (filtering) cycle of the Kalman filter to the linearized observation dynamics to obtain $\hat{\mathbf{X}}_{k+1|k+1}$ and $\mathbf{P}_{k+1|k+1}$.

Since the EKF is not an optimal filter, $\mathbf{P}_{k+1|k+1}$ and $\mathbf{P}_{k+1|k}$ do not represent the true covariance of the state estimates as they do with the standard Kalman filter. If the observation function is linear, then the corrected state and error-covariance matrices are found as before, using Eqs. (5.36) through (5.39).

5.14 Summary

Kalman filtering is probably the best-known technique for updating the state and error-covariance estimates corresponding to the track of an object. It computes optimal, in the least-squares estimate sense, corrected (or updated or filtered or *a posteriori*) state estimates and state error-covariance estimates at time step k given a measurement at time step k. It also provides a mechanism for projecting the state and error-covariance estimates forward to time step $k+1$ as one-step-ahead predictions or *a priori* estimates. When the measurement noise is generated by taking random samples from the noise distribution, the consistency of the filter initialization is guaranteed. If several

Monte Carlo runs are made, random sampling of the noise distribution is performed for each run so that new and independent noise values are incorporated into every run.

Practical implementation of a Kalman filter requires knowledge of the equations that govern evolution of the state of the tracked object with time, and the ability to add process noise of sufficient value to prevent the Kalman gain from becoming negligible. If the latter were to occur, the tracker would simply dead recon the future position of an object and ignore current and future measurements. On the other hand, adding excessive amounts of process noise most likely will result in a clutter-to-measurement correlation and cause the tracker to lose the *bona fide* target.

An IMM that uses four models of vehicle motion to represent common vehicle-tracking scenarios was described to illustrate the application of Kalman filtering to estimate a vehicle's trajectory 3 s into the future. The transition probability matrix that applies to this particular problem was presented to show the reader an example of its structure. The IMM estimate was improved by using GPS location data to correct an inaccurate state estimate, particularly on curves where the IMM often placed the vehicle outside of the road boundaries.

When the system dynamics are nonlinear, the extended Kalman filter may be used to linearize the motion about the current mean and covariance of the state through first-order Taylor approximations to the time-varying transition and observation matrices. Since the EKF is not an optimal filter, the error-covariance values do not represent the true covariance of the state estimates as they do with the standard Kalman filter. There are alternatives to the EKF when the system of interest is highly nonlinear, as discussed in Section 2.7.

References

1. A. Gelb, J. F. Kasper, Jr., R. A. Nash, Jr., C. F. Price, and A. A. Sutherland, Jr., *Applied Optimal Estimation*, M.I.T. Press, Cambridge, Massachusetts (1979).
2. P. S. Maybeck, *Stochastic Models, Estimation, and Control, Volume 1*, Academic Press, New York (1979).
3. S. S. Blackman, *Multiple-Target Tracking with Radar Applications*, Artech House, Deedham, Massachusetts (1986).
4. S. S. Blackman, "Association and fusion of multiple sensor data," Chapter 6 in *Multitarget-Multisensor Tracking: Advanced Applications*, Y. Bar-Shalom, Ed., Artech House, Norwood, Massachusetts (1990).
5. G. Welch and G. Bishop, *An Introduction to the Kalman Filter*, TR 95-041, Dept. of Computer Science, Univ. of North Carolina at Chapel Hill, July 2006. Also available at www.cs.unc.edu/~welch/kalman/kalmanIntro.html. Accessed Jul. 14, 2011.

6. M. S. Grewal and A. P. Andrews, *Kalman Filtering Theory and Practice*, Prentice Hall, Upper Saddle River, New Jersey (1993).
7. Y. Bar-Shalom and T. E. Fortmann, *Tracking and Data Association, Mathematics in Science and Engineering* **179**, Academic Press, Orlando (1988).
8. Y. Bar-Shalom, P. K. Willett, and X. Tian, *Tracking and Data Fusion: A Handbook of Algorithms*, YBS Publishing, Storrs, Connecticut (2011).
9. "State Space Models and the Kalman Filter," *Computational and Mathematical Engineering (CME)* **308** Notes, Chapter 7, Stanford University, Stanford, California (2011).
10. X. R. Li and V. P. Jilkov, "Survey of maneuvering target tracking—Part III: Measurement models," *Proc. SPIE* **4473**, 423–446 (2001) [doi: 10.1117/12.492752].
11. X. R. Li and V. P. Jilkov, "A survey of maneuvering target tracking—Part IV: Decision-based methods," *Proc. SPIE* **4728**, 511–534 (2002) [doi: 10.1117/12.478535].
12. R. A. Singer, "Estimating optimal tracking filter performance for manned maneuvering targets," *IEEE Trans. on Aerospace and Electronic Systems* **6**(4), 473–483 (1970).
13. X. R. Li and V. P. Jilkov, "Survey of maneuvering target tracking—Part I: Dynamic models," *IEEE Trans. on Aerospace and Electronic Systems* **39**(4), 1333–1364 (2003).
14. M. I. Ribeiro, *Kalman and Extended Kalman Filters: Concept, Derivation and Properties*, Institute for Systems and Robotics, Lisbon (2004). omni.isr.ist.utl.pt/~mir/pub/kalman.pdf. Accessed Jul. 29, 2011.
15. R. J. Schwarz and B. Friedland, *Linear Systems*, McGraw-Hill, New York (1965).
16. H. F. Durrant-Whyte, *Introduction to Estimation and the Kalman Filter* V2.2 (2001). www.isip40.it/resources/Dispense/DataFusionSIIT/Estimation_KF.pdf. Accessed Aug. 6, 2018.
17. M. P. Dana, *Introduction to Multi-Target, Multi-Sensor Data Fusion Techniques for Detection, Identification, and Tracking, Part II*, Johns Hopkins University, Organizational Effectiveness Institute Short Course ROO-407, Washington, D.C. (1999).
18. M. P. Dana, "Registration: A prerequisite for multiple sensor tracking," Chapter 5 in *Multitarget-Multisensor Tracking: Advanced Applications*, Y. Bar-Shalom, Ed., Artech House, Norwood, Massachusetts (1990).
19. J. E. Gray, "What a long strange trip it has been: My twenty-five year affair with the alpha-beta filter," engineering.jameskbeard.com/For_-Download/filterbionf.pdf. Accessed Jul. 24, 2018.
20. P. Kalata, "α–β target tracking systems: A survey," 1992 *IEEE American Control Conference Proceedings*, June 24–26, Chicago, pp. 832–836 (1992).

21. S. Hong, C. Lee, F. Borrelli, and J. K. Hedrick, "A novel approach for vehicle inertial parameter identification using a dual Kalman filter," *IEEE Trans. on Intelligent Transportation Systems* **16**(1), 151–161 (2015).
22. C. Hu, W. Chen, Y. Chen, and D. Liu, "Adaptive Kalman filtering for vehicle navigation," *Journal of Global Positioning Systems* **2**(1), 42–47 (2003).
23. M. Schreier, V. Willert, and J. Adamy, "An integrated approach to maneuver-based trajectory prediction and criticality assessment in arbitrary road environments," *IEEE Trans. on Intelligent Transportation Systems* **17**(10), 2751–2766 (2016).
24. C. Barrios and Y. Motai, "Improving estimation of vehicle's trajectory using the latest global positioning system with Kalman filtering," *IEEE Trans. on Instrumentation and Measurement* **60** (12), 374–3755 (2011).
25. Y. Bar-Shalom, K. C. Chang, and H. A. P. Blom, "Automatic track formation in clutter with a recursive algorithm," Chapter 2 in *Multitarget-Multisensor Tracking: Advanced Applications*, Y. Bar-Shalom, Ed., Artech House, Norwood, Massachusetts (1990).
26. X. R. Li and V. P. Jilkov, "Survey of maneuvering target tracking—Part V: Multiple model methods," *IEEE Trans. on Aerospace and Electronic Systems* **41**(4), 1255–1321 (2005).
27. B. Anderson and J. Moore, *Optimal Filtering*, Prentice-Hall, Englewood Cliffs, New Jersey, pp. 195–204 (1979).

Chapter 6
State of the Practice and Research Gaps

6.1 Data Fusion State of the Practice

Methods and standards for implementation of fusion systems are evolving as research continues concerning the nature and enhancement of the human–computer interface. Architecture selection, implementation, and test processes are still *ad hoc*, often driven by communication and data-processing constraints, and frequently dictated by personal taste and corporate and agency culture.

Advances in processor technology and sensor netting techniques have removed many of the limitations of the past. Improved signal-processing techniques and digital sensor technology have reduced clutter and false-alarm issues in some circumstances. Improved workstations, user menu interfaces, and continued exploration of the function of the user in the data fusion process have accommodated many user needs and broadened the understanding of human behavior patterns stemming from access to fused data. Many of these topics were discussed in Chapter 2.

However, operational limitations of commercial-off-the-shelf (COTS) hardware and software may inhibit the full use of new data-processing innovations. Commercial operating systems and database management systems (DBMS) may be ill-suited to some ITS applications because they often cannot support real-time requirements for sensor data processing. The introduction of connected, cooperative, and self-driving vehicles requires that DBMS be designed for worst-case data access requirements, as data transmission and processing delays in safety-critical applications are unacceptable.

In state estimation, data correlation is the largest user of data-processing resources, often more than 60 to 70% of the total.[1] Techniques for computer performance modeling are still primitive. Detailed transaction analysis is required as an input. Operating systems, DBMS, and other support functions have historically not been included in the modeling or analysis. Scaling up from simple situations underestimates the data-processing requirement, particularly for multiple-hypothesis tracking techniques. To date, rapid prototyping proves to be the best solution for estimating data-processing requirements.

Data transmission and processing requirements for ITS systems should be specified for worst-case scenarios, such as in connected vehicle and self-driving vehicle applications where data access and transmission delays (latency) may affect the safety of travelers. This fact was appreciated in a 2016 proposed rule making by the National Highway Traffic Safety Administration (NHTSA) that specified the designated short-range communications (DSRC) protocol as the primary technology for *safety-critical* V2V communications.[2] The latency of 200 μs for DSRC is two orders of magnitude less than the most stringent latency requirement of 0.02 s needed for connected-vehicle active-safety applications. Although 5G networks promise latency response times of 1 ms, they currently lack the necessary infrastructure and related telematics.[3]

Notwithstanding the U.S. Department of Transportation's and NHTSA's delay in pursuing the rule making since then, several automobile manufacturers have decided to introduce DSRC, which is based on IEEE 802.11p, into their vehicles beginning in the early 2020s. Among the manufacturers are Toyota, Lexus, General Motors, Volkswagen, and Renault. In fact, a recent study by the University of Michigan Transportation Research Institute notes that a three-year delay in the deployment of V2V systems will allow the sale of millions of new cars and light trucks without them.[4] Since those vehicles will remain on the road for an average of 15 years, they would have millions of avoidable collisions during that period. Furthermore, installing DSRC-based systems now will save 40,000 lives in the U.S., another 40,000 in western Europe, and still more around the world.

The platform for the deployment of Cooperative Intelligent Transport Systems in the European Union (EU) (known as the C-ITS Platform) contains an alternative V2V communications proposal for vehicles manufactured in the EU.[5] Its hybrid communications concept would use the immediately available IEEE 802.11p/ETSI ITS-G5 communications protocol for short-range communications in the 5.9-GHz band for the services it could support, with further study to investigate whether geographical coverage and other cooperative ITS services can be increased through the cellular infrastructure. A summary of the C-ITS Platform's Final Phase I Report is found in Ref. 17. In addition to the above proposal, Regulation (EU) Number 165/2014 of the European Parliament requires trucks carrying loads greater than some specified mass and registered after June 2019 to have a DSRC interface that enables remote reading of tachograph data.

6.2 Need for Continued Data Fusion Research

There are ongoing needs for data fusion research in many areas, including:

- Reliability and quality of input data to the fusion system,
- Security of the data fusion system,

- Fusion of hard and soft data,
- Assessing the fusion system using measures of performance (MoPs),
- Adaptive nature of MoPs,
- MoP dependence on operational scenario,
- Measures of effectiveness (MoEs) for information fusion,
- Ground truth,
- Commercial database management and operating systems suitability, and
- Designing for worst-case data transmission and processing scenarios.

6.2.1 Reliability and quality of input data to the fusion system

Approaches are needed to assess the fusion system input data and to calculate the degree of confidence in the data in terms of their reliability and credibility.[6–8] Khaleghi, Khamis, Karray, and Razavi[9] state that the most notable work in this area are the standardization agreements (STANAGs) of the North Atlantic Treaty Organization (NATO).[10] Here an alphanumeric rating system combines a measurement of the reliability of the information source with a measurement of its credibility, where both measures are evaluated using the existing knowledge. STANAG recommendations are expressed using natural language statements, making them imprecise and ambiguous. The proposed formalism is based on the observation that three notions are required in an information evaluation system: the number of independent sources supporting the information, their reliability, and knowledge that the information may conflict with some available or prior information.

Cholvy incorporates the notion of degree of conflict, in contrast to merely conflicting or nonconflicting information.[11] Nonetheless, the current formalism is not complete as there are some aspects of the STANAG recommendations, such as total ignorance about the reliability of the information source, that are not considered.

The rate at which information is provided to the fusion system is another critical aspect of input information quality. The information rate is a function of several factors, including the revisit rate of the sensors and other data sources, the speed at which data sets are communicated, and the quality of the communication link.[12] The effect of the information rate is particularly significant in decentralized fusion architectures where imperfect communication is common.

Uncertainties in data interpretation arise not only from imprecise and noisy sensor measurements, but also from the inability to distinguish between ambiguities and inconsistencies in the environment or scenario as reported by the sensors and opportunistic data sources (e.g., cell phone reports, connected vehicle transmissions, and highway patrol reports). Examples of "uncertain data" are outliers and spurious data.[13] For example, when measuring travel times on a highway using vehicle reidentification techniques (e.g., toll tag,

Bluetooth® MAC address, and automatic license-plate readers; trucking industry transponders), data from vehicles stopping at rest areas can convey false information in the form of increased travel times, while motorcycles weaving between lanes can do just the opposite.[14] A different issue arises when traffic management personnel use the number of measurements reporting the same value as a measure of the accuracy or reliability of the data. Such an approach begins to fail when the roadside readers reidentify a large number of cell phones on a bus. Although the travel-time data match each other, they are not independent measures of the parameter of interest. A variation of this concern appears when a pedestrian who is walking is reidentified by a MAC address reader several times over some number of minutes. In this case, the travel-time measuring device is not in a vehicle at all and the multiple acquisitions of the information should be filtered by an appropriate means.[15]

6.2.2 Security of the data fusion system

Whether military or transportation related, data fusion applications require data integrity, confidentiality, and freshness. Preliminary efforts notwithstanding, the security aspect of data fusion systems is still largely unexplored, especially in large-scale sensor networks covering sizable areas and in support of MoEs that rely on safe and protected data permission credentials and access.[9,16]

Security and credentials management (SCM) is an integral part of the U.S. Connected Vehicle Program and similar efforts underway in the EU. These programs combine data received from vehicles, infrastructure communications devices, and pedestrians in their vicinity. SCM is a set of support applications that ensure trusted communications between one mobile device and another or between a mobile device and a roadside device, and ensure the protection of data under their control from unauthorized access. The applications allow credentials to be requested and revoked, and secure the exchange of trust credentials between parties so that no other party can intercept and use those credentials illegitimately.[17]

6.2.2.1 Public key infrastructure systems

Research indicates that a public key infrastructure (PKI) security system involving the exchange of digital certificates among trusted users can support the need for both message security and appropriate anonymity to users.[18] Digital certificates are used to sign the messages that pass between vehicles in the connected vehicle environment and therefore allow the receiver of a message to verify that the message came from a legitimate source. The receiver of the message also needs to confirm that the sender has the correct certificate to not only send the message, but also to give the commands it contains.

PKI systems rely on encryption for security. Symmetric encryption systems use one key both to encrypt and decrypt the message. In an asymmetric encryption system such as PKI, keys come in pairs—the sent message contains one half of the key pair, while the receiving device has the other half—potentially enabling a PKI system to be more secure. Additional information concerning PKI systems and security systems proposed for V2V communications is found in Chapter 12 of Klein.[17]

6.2.2.2 Limitations of existing PKI systems

Today's off-the-shelf PKI systems are not broad enough in their functions to serve as the crucial safety-critical element for V2V communications. Most PKI systems are concerned with data exchange among parties that are either known to each other as trusted sources (e.g., the military knows each of its communication points) or are identifiable (e.g., air traffic controllers can identify each of the planes involved in safety-critical data exchange). Also, the majority of other safety-critical systems employ highly secure networks or private networks (e.g., the military) and cannot leverage either existing communications systems or the Internet (to keep capital investment costs to a minimum and to achieve widespread access) in a manner that does not introduce additional vulnerabilities and risks.[19]

Nearly all of the existing commercial systems, by comparison, do leverage the Internet and wireless systems. These systems enable online purchasing or online financial transactions in a manner that allows for easy accessibility to millions of users. They do not, however, meet the level of privacy protection required for V2V data exchanges, as these organizations have pre-existing agreements with the certificate authority (CA). Consequently, user identity resides within databases and is typically part of the authentication process. The CA is an entity that acts as the trusted third party to provide the action to authenticate the entities within a network. It normally does so by signing and distributing digital certificates. To protect privacy, these short-term certificates contain no information about users, but serve as credentials that permit them to participate in the V2V system. The CA also typically revokes certificates and publishes a certificate revocation list (CRL) so that valid users know to ignore certificates of users that have been revoked. A CA is considered the root of trust in a PKI.[18]

6.2.3 Fusion of hard and soft data

Fusion of hard and soft data is of special importance in situation awareness applications. However, differences between hard or sensor data and soft data (e.g., human-sourced data, whether voiced or written, or from a database) make heterogeneous fusion a challenge. According to Dragos and Gatepaille,[20] three crucial distinctions exist:

1. Sensor data are typically delivered in continuous streams, often at well-defined time intervals without prodding. Those streams need to be processed in near real time because they often reveal a critical event, such as a traffic accident, through information that requires an immediate response. In contrast, soft data arrive sporadically, sometime after the event occurred.
2. Sensors do not alter the delivered data in a manner that changes the data's value or implications. In contrast, soft data are the result of the sender's perception, affect, skills, and knowledge as influenced by subjective assessments due to social and cultural experiences that shape the sender. Furthermore, humans have the ability to provide a personal view, attitude, or appraisal in the form of a judgment or belief when reporting the details of an event.
3. Sensors are typically connected in *ad hoc* networks that cover a geographic area. This permits the user receiving data from arbitrary nodes to consider network effects, if any, when performing the analysis. Unlike sensors, human sources can be part of hidden networks whose undisclosed ties can lead to unreliable conclusions. For example, dependent pieces of information transmitted by cell phone conversations from several different vehicles traveling closely together (or from passengers on the same bus) can be analyzed under the false assumption that independent sources exist, when in fact the sources may not be independent at all.

6.2.3.1 Attributes of hard and soft data

Hard and soft data are used to identify features that describe entities, which in a traffic management scenario can be individual vehicles, vehicle platoons, bicyclists, pedestrians, animals, or roadside objects. An entity's feature vector, when fused with other feature vectors, gives the position and kinematics of the entity, its type, and its relation to contextual information such as geographical features (roads, waterways, forests, deserts, other landmarks) or to other entities in the scenario. Entity states can be built upon sensor-based data and soft observation reports. The core of the Dragos and Gatepaille framework[20] for a homeland defense or military scenario is a sensor-based kernel fusion that supports several processes for entity correlation and tracking and estimation of their states. Sensor data features are heterogeneous due to the different types of deployed sensors. Sensor data arrive as formatted reports integrating position, time, type, and potential subtype of an observed entity. Additional features may be present such as vehicle color or vehicle identification code (sometimes available from transponders located on the vehicle). The second layer of their architecture enriches the previously created situation by integrating soft data elements when they arrive on an irregular basis. Soft data observations are contained in structured reports with natural language paragraphs that summarize information collected by human sources.

6.2.3.2 Analyzing soft data

Processing of soft data proceeds by identifying properties of entities within natural language messages whose size can range from a few phrases to several sentences. Messages also have a heterogeneous content and can provide insights leading to entity location, evolution, and situation. The attribute values extracted from messages can then be modeled as properties of the entities and integrated into entity states. Linguistic analysis reprocesses documents to split texts and performs lexical normalization that removes the same type of information provided by different lexical forms. Linguistic analysis also identifies sentence boundaries and converts a sequence of characters into a sequence of tokens (tokenizing) while filtering a list of stop words. Tokens (examples of which are regular expressions, specific sequences of characters, specific separating characters, and explicit definitions from a dictionary) are identified based on the specific rules used by the individual performing this function. Stop words are those that appear to be of little value in helping select text matching a user's need. Examples of these words are *a*, *an*, *the*, *is*, *at*, *which*, *on*, *be*, *by*, *for*, *it*, *of*, *was*, and *to*. This phase of soft data analysis produces a set of transactions, where each transaction is a sentence whose items are words. The set of transactions is then used to extract binary collocations.[20]

Collocations are associations of words that co-occur frequently within the same sentence, whether because the meanings of the words are related to each other (such as vehicle–road, car–driver, inclement–weather, injury–accident) or because the two words make up a compound noun (e.g., rest area, bus stop, train station, travel time). Attributes are identified from soft data by utilizing a text mining approach centered on identifying collocations composed of two words, also named by-grams, as entities of interest are often named by short lexical units. Because simply taking the entire list of collocations captures an excess of extraneous and incoherent information, additional processing is necessary to filter relevant word associations. Semantic annotation is performed automatically by using procedures based on lexical similarities that associate a real number to a pair of words and offers a measure of the similarity between two words. Lexical similarities are used to label collocation by ontological concepts.

Situation enhancement occurs through two mechanisms. First, complementary properties extracted from human reports are added to the states of entities inferred from hard data. Then the entity type is updated by considering the types derived from sensor or hard data processing and the types extracted from soft data.[20] Assignment of human observations to the entities in a particular situation is performed by spatiotemporal correlation. As observations are associated with a time stamp and have specific locations, this method first estimates a correlation coefficient to describe the probability of a human observation to be assigned to E_i, an entity belonging to the situation. Reasoning mechanisms that combine attributes of entities are used to update the type; e.g., the label "vehicle" may be changed to "bus," "SUV," or "18-wheeler." The results of this estimation are then ranked, and the observation maximizing the value is selected and added to the set of observations associated with E_i.

The reliability of human reports is not guaranteed as they can be noisy, incomplete, and at times irrelevant. In these circumstances, the inference mechanism may fail to identify the most specific type label, and the fusion result is inconclusive. Inconclusive inference highlights contradictions between sensor reports and observations or accidental associations of human reports to erroneous entities. When successful, the result of fusing soft and sensor data at the entity level is a more accurate identification of the entity type, along with the enrichment of their state feature vector by the properties contributed by the soft data.

6.2.4 Assessing the fusion system using measures of performance

Performance evaluation has different and possibly conflicting facets that may be difficult to capture in one comprehensive and unified measure. One can argue that performance assessment should be multifaceted, not only measuring how well fusion goals were achieved but also the amount of effort and resources spent to accomplish this task. Therefore, a comprehensive MoP might become too abstract and fail to properly reveal all dimensions of system performance. The alternative is to deploy a set of MoPs tailored to the application.[9] Although it may be difficult to devise a unified MoP to capture all aspects of system performance in a comprehensive manner, specific applications may lend themselves to this approach, as described in Section 6.2.4.3.

6.2.4.1 Adaptive nature of MoPs

The more complex the scenario, the more challenging it becomes for the fusion system to maintain the desired performance level. Thus, a fair indicator of fusion performance may require evaluation criteria that adapt over time or according to the given context and situation.

6.2.4.2 MoP dependence on number of objects of interest

The selection of specific MoPs depends on whether the scenario contains one or many objects of interest. For example, a system with multiple objects of interest consists of more than one feature vector and, hence, is fundamentally different from a single-object system involving a single feature vector. This characteristic of multi-object systems is due to the appearance and disappearance of the objects (e.g., numbers and types of vehicles, bicyclists, and pedestrians) over time, which causes the number of states to vary with time. Likewise, multiple-sensor systems have more parameters to evaluate than single-sensor systems. One such example is the assessment of the data and track association and estimation processes that are integral to the multi-object detection and tracking problem.

Metrics can be computed for individual objects or for an ensemble of objects. Individual object metrics for the state estimation application include track accuracy, track covariance consistency, track jitter, track estimate bias, and track continuity. Ensemble object metrics include average number of missed objects, average number of extra or false objects, average track initiation time, completeness history, and cross-platform commonality history.[9]

6.2.4.3 MoEs for information fusion

Waltz and Llinas,[21] Blasch, Valin, and Bosse,[22] and Blasch, Breton, and Valin[16] differentiate between MoPs and MoEs. These authors view MoEs as combinations of one or more MoPs that are dependent on the application. In military applications, MoPs are related to physical and structural dimensional parameters that measure attributes of system behavior. Examples that have crossover to the transportation field are detection probability, false-alarm rate, accuracy of object location estimate, object identification or classification probability as a function of range, communication latency period, and sensor spatial coverage. Effectiveness is concerned with three issues: (1) efficiency—executing jobs in the most economical way, (2) efficacy—getting assignments accomplished in a manner that meets objectives, and (3) effectiveness—performing the tasks required to achieve an overall goal.[22]

Thus, MoEs gage how well a command, control, and communications system executes its functions within an operational environment. They are dependent on the sensor types, environmental conditions, objects of interest, situational context, and user needs.[16] Pertinent characteristics for ITS are timeliness, accuracy, and availability of data and information; incident warning time; and ability to function in inclement weather and in a variety of road configurations, vehicle mix, and traffic flow rates.

Blasch, Valin, and Bosse[22] and Blasch, Breton, and Valin[16] introduce information gain, information quality, and robustness as MoEs suitable for information fusion and decision support systems. These MoEs are based on MoPs of timeliness, accuracy, confidence, throughput, and cost. Information gain includes assessments of the fusion products (e.g., confidence in the results of the analysis) and usability in the form of information completeness and appropriateness of the interface design.[16] Information quality depends on the reliability or timeliness of service and data and source credibility. Robustness is an MoE that incorporates the context and number of use cases or scenarios that can be captured.

Dragos and Gatepaille[20] compute information gain (IG) based on the premise that changing the state of entities (i.e., the objects of interest) by incorporating observations improves the description of the entities. Accordingly, IG is found as

$$\mathrm{IG} = \frac{N_c}{N}, \tag{6.1}$$

where N_c is the number of entities whose states are affected by observations, and N is the overall number of entities in the scenario. IG has values between 0 and 1. The limiting value of zero occurs when observations are not related to the entities of the situation and when states of all entities are updated under ideal conditions.

Information quality (IQ) accounts for the number of failures due to inconclusive fusion as[16]

$$\mathrm{IQ} = 1 - \frac{U}{N_c}, \tag{6.2}$$

where U is the number of inconclusive inferences. IQ extends between 0 (when all inferences for type refinement are inconclusive) and 1 (when all are valid).

Robustness R quantifies the consistency over the testing and application domains, i.e., variations that occur in real-world applications. When a system is able to function with real-time perturbations in its operating scenario, robustness R equals 1.

6.2.4.4 MoEs for risk management

Falcon, Abielmona, Billings, et al.[23] explore MoEs that are useful in a risk management framework (RMF) architecture that characterizes and evaluates maritime vessels in potential distress. An extension to ITS is suggested in the last paragraph of this section. Risk features pertaining to maritime situation awareness are defined *a priori* and then quantified in real time using both hard and soft data. Hard data include periodic reports of latitude, longitude, and vessel type from automatic identification systems (AIS); radar ship position data; video camera images; and weather gauge data. Examples of soft data are historical reports of worldwide maritime incidents and geographical databases that contain textual repositories of city and location names and their respective geographical coordinates that supplement incident reports lacking position information.

Falcon, Abielmona, Billings, et al.[23] utilize the general information fusion effectiveness (IFE) MoE developed by Blasch, Valin, and Bosse[22] that contains three multiplicative factors: information gain, information quality, and robustness. However, they inject explicit risk functions—risk awareness level (RAL) and overall information timeliness (OIT)—into the first two factors such that

$$\mathrm{IFE} = \underbrace{\mathrm{IBR} \times \mathrm{RAL}}_{\substack{\text{Information}\\ \text{Gain}}} \times \underbrace{\mathrm{IC} \times \mathrm{OIT}}_{\substack{\text{Information}\\ \text{Quality}}} \times \underbrace{\mathrm{R}}_{\text{Robustness}}. \tag{6.3}$$

In the maritime situation awareness scenario, information gain is the ratio of the numbers of vessels that use the available hard and soft data sources to the total number of vessels in the scenario [the information benefit ratio (IBR)] multiplied by the overall risk alertness they provide (the RAL). The RAL is defined as the extent to which the vessels in the scenario are aware of the existing risks. This variable is computed as the average overall risk, across all risk features, of the vessels in the scenario at any time instant.

Information quality is defined as the product of the information confidence (IC) and the OIT. Information confidence is calculated as the mean of the product of the reliability of information (ROI) and the reliability of the source (ROS) for all participating hard and soft data sources. Accordingly, IC takes the form

$$\mathrm{IC_{HS}} = \frac{\sum_{i=1}^{m} \mathrm{ROS}_{\mathrm{H}_i}\, \mathrm{ROI}_{\mathrm{H}_i} + \sum_{j=1}^{n} \mathrm{ROS}_{\mathrm{S}_j}\, \mathrm{ROI}_{\mathrm{S}_j}}{(m+n)}, \tag{6.4}$$

where the subscripts H and S represent hard and soft data sources, respectively, of which there can be more than one in each category.

OIT is a function of information timeliness (IT), which is composed of two terms, one for hard data and another for soft. $\mathrm{IT_H}$ is expressed as the average delay (in seconds) between consecutive AIS reports for each vessel averaged across all vessels. $\mathrm{IT_S}$ is the average age (in days) of each vessel's n closest maritime incident reports averaged across all vessels. In the model developed by Falcon et al., $\mathrm{IT_H}$ and $\mathrm{IT_S}$ are the crisp outputs of a fuzzy logic process that evaluates the values of hard-information timeliness (HIT) and soft-information timeliness (SIT), respectively. Defuzzification is performed by a center-of-gravity computation. Robustness is set to 1 since the system can function with real-time variations.[23]

This IFE MoE can be extended to ITS in a variety of ways: by considering vehicles rather than vessels in the architecture; using studies to evaluate the IBR as a function of vehicle location (by geographical location in a country, state, county, or city; by roadway type and by lane—Interstate, urban, rural, unimproved; by congestion level; etc.) and the types or population of vehicles in use at a particular time; evaluating drivers' RAL, perhaps with surveys; investigating the reliability of ROI and ROS as they pertain to the various sources of traffic data described in Section 2.3; adapting the IT_H and IT_S concepts by considering the delays attributable to the hard and soft traffic data sources, respectively, as they affect drivers, other travelers (e.g., transit users), traffic management personal, and other stakeholders; and incorporating additional suitable analogies.

6.2.5 Ground truth

Ground truth is not usually available, yet many of the currently used MoPs require knowledge of ground truth values. This is the most common and serious issue that still needs to be addressed according to Khaleghi, Khamis, Karray, and Razavi.[9] A potential solution is to develop objective performance measures, i.e., measures that are independent of ground truth or human subjective evaluation. Some progress has been made in the image fusion community using metrics such as those described in Ref. 24 and in the information fusion community with the metrics discussed above.

6.2.6 Commercial database management system and operating system suitability

Database management software accounts for the largest amount of software in a data fusion system, even if one utilizes sophisticated COTS DBMS packages.[25] Fusion system inputs originate from sensor data (e.g., scalars, vectors, time series, images), information contributed by human users (e.g., cell phone calls, incident reports), environmental data, textual information, and knowledge such as road configurations, planned events, and local driving culture. Hence, the use of existing DBMS for data fusion may pose concerns for the following reasons:

- Existence of commercial DBMS that may be ill suited to ITS real-time requirements for sensor data processing and control of critical safety-related functions as they are designed for flexibility of application rather than real-time or fast-time processing.[26] This consideration also often applies to commercial operating systems.
- Existence of large and varied databases with numerous records and record formats. Consider the collection, storage, and processing requirements imposed by big data acquired through crowdsourcing from connected vehicles and cell phone location and tracking devices. Gettman, Toppen, Hales, et al. have estimated that the storage required in a typical traffic management agency for emerging data from connected travelers, connected vehicles, and new types of connected infrastructure could be in the many thousands of terabytes by 2026.[27] References 27 and 28 provide information concerning big data tools and technologies for aggregating, storing, and analyzing data from connected travelers, connected vehicles, connected

infrastructure, and other emerging data sources including systems that monitor: lane volumes and speeds (and adjust speed limits accordingly), parking availability, emissions, weather, road condition and performance, and transactional data for agencies involved with Transportation Systems Management and Operations (TSM&O), such as traffic management centers and service providers. An introduction to big data as it affects managing large data sets used for traffic monitoring is found in Ref. 29.

- Requirement to simultaneously accept data at a rate governed by the contributing sensors and other sources, while allowing automated fusion processes (such as data association) and users to rapidly retrieve large amounts of data using, for example, general Boolean queries. The combination of data set complexity and the need for real-time data storage and retrieval complicate database management for data fusion.[25]
- Maintenance of data integrity in real time under rapid receipt of sensor data; intense human interactions; asynchronous, out-of-sequence, and false sensor reports; etc.
- Need to provide flexible and user-friendly interfaces.
- Acceptance of both fixed format and free-text message formats under multiple protocols.
- Recognition that data associated with fusion systems may involve security when police reports are involved.

In addition, DBMS are affected by data maintenance that impacts reliability, survivability, and supportability. Reliability is concerned with making the data available, correct, and timely, implying that the data should be stored and accessed in such a way that they can meet these requirements. Survivability involves collecting and correlating data to enable their effective insertion into the intended application. Survivability is supported by ensuring that data have the correct security labels and are annotated with space, time, and event notations. Supportability is sustained by ensuring that the data have the associated contextual metadata information needed to register, exploit, and fuse the data with other sources.[16]

6.2.7 Design for worst-case data transmission and processing scenarios

ITS operations should be designed for worst-case scenarios such as incidents that occur in automated driving applications that involve safety where delays at critical times are unacceptable. Rapid prototyping is the best solution for estimating data-processing requirements. Guidelines for rapid prototyping include using the target machine if possible, using prototype software in the required language, and driving the analysis with the worst-case load.[30]

6.2.8 Additional research needs

Data fusion applications to transportation and traffic management have been ongoing for at least two decades and have produced many successful applications, especially in the low-level fusion processes (Level 0 and Level 1). However, these encouraging results should not conceal the challenges that remain before deployment of operational data fusion systems, particularly those exploiting the higher-level fusion processes, becomes the norm in transportation management. Challenges include (1) obtaining data with the necessary accuracy and availability to make fusion effective, (2) dynamic and real-time concerns associated with data quality as traffic flow changes, (3) processing the data in real time, and (4) finding methods to effectively combine sensor or hard data with human-generated or soft data.[9,31] The last item is particularly critical for situation awareness analyses. Of key interest to highway traffic monitoring personnel is a desire to combine different sources of travel-time data to improve the reliability of the estimate and extend the region over which the predictions are available. Another goal is to develop open-source analysis tools to process big data such as speed and travel-time data from conventional sources and crowdsourcing.[27]

Prospects for further road transportation data fusion applications appear promising due to the increased collection of data from sources other than roadside sensors, such as those identified in Section 2.3. Wireless technologies that offer (1) the potential of easier reporting and access to customized information (e.g., cooperative systems with V2X and pedestrian-to-vehicle or infrastructure communications) and (2) the ability to track individual vehicles to obtain floating car data (FCD) and extended FCD will enrich the information pool and accelerate desires for operational data fusion systems, not only for low-level processing, but also for higher-level processing functions. This is already appearing in England through the National Traffic Information Service, as mentioned in Section 2.14.

Despite the noted benefits of data fusion, issues remain that may potentially delay access to connected vehicle and pedestrian data. These include standardizing security and credentials management procedures that ensure secure and trusted communications between vehicles and devices carried by pedestrians or between vehicles or pedestrians and roadside or other infrastructure devices, finalizing the V2V and V2I communications technology used for safety-critical connected vehicle applications and the timetable for its appearance in new vehicles, technical policy issues (e.g., software safety verification and validation methods), legal policy issues (e.g., insurance, liability, and risk management options; intellectual property, data ownership, and privacy policies), implementation policy issues (e.g., agreeing on governance structures that identify roles and responsibilities for the various participants), implementation strategy issues such as identifying funding sources for required infrastructure improvements, and institutional issues that arise when the stakeholder community consists of multiple agencies (e.g., defining an interagency concept of operations, harmonizing agency contracting and procurement practices, identifying funding for operations and maintenance, and ensuring the availability of staff with the necessary technical skills).[17]

6.3 Prerequisite Information for Level 1 Object Assessment Algorithms

Many of the Level 1 data fusion algorithms are mature in the context of mathematical development. These algorithms encompass a broad range from numerical techniques to heuristic approaches such as clustering and knowledge-based expert systems. Algorithm selection criteria and the requisite *a priori* data, however, are still major challenges as can be inferred from the discussions found in the preceding chapters.

Applying Bayesian inference, Dempster–Shafer evidential reasoning, artificial neural networks, voting logic, fuzzy logic, and Kalman filtering data fusion algorithms to object detection, classification, identification, and state estimation requires expert knowledge, probabilities, or other data and information from the designer to define one or more of the following groups of factors:

- *A priori* probabilities and likelihood functions (Bayesian inference),
- Probability mass (Dempster–Shafer evidential reasoning),
- Neural-network type, numbers of hidden layers and weights, and training data sets (artificial neural networks),
- Confidence levels and conditional probabilities (voting logic),
- Fuzzy set state variables, membership functions, production rules, and defuzzification method (fuzzy logic), and
- Kinematic and measurement models for the objects of interest, process noise, and model transition probabilities when multiple state models are utilized (Kalman filtering).

Table 6.1 summarizes the prerequisite data and information needed to implement the data fusion algorithms that were discussed.[30] Data fusion algorithm selection and execution is dependent on the expertise and knowledge of the specialist who designs the data fusion concept (e.g., to identify the data fusion procedure and the methods for obtaining the required data fusion algorithm inputs), analysis of the operational situation (e.g., to identify objects of concern and how their states evolve with time, available information sources, weather and other operational constraints, special events), applicable information stored in databases (e.g., to calculate the required prior probabilities, likelihood functions, or confidence levels), types of information provided by the sensor data or readily calculated from them (e.g., to calculate likelihood functions, probability masses, or to differentiate

Table 6.1 Data and information needed to apply Bayesian inference, Dempster–Shafer evidential reasoning, artificial neural networks, voting logic, fuzzy logic, and Kalman filtering data fusion algorithms to object detection, classification, identification, and state estimation.

Data Fusion Algorithm	Required Data and Information	Example
Bayesian inference	*a priori* probabilities $P(H_i)$ that the mutually exclusive hypotheses H_i are true.	Using archived sensor data or sensor data obtained from experiments designed to establish the *a priori* probabilities for the particular scenario of interest, compute the probability of detecting a desired object given that data are received by the sensor. The *a priori* probabilities are dependent on the scenario of interest and take into account such factors as the numbers of the different types of objects that are present and the size of the surveilled area. If *a priori* probabilities cannot be established, invoke the principle of indifference to obtain their initial values.
	Likelihood probabilities $P(E\|H_i)$ of observing evidence E given that H_i is true as computed from experimental data from conditionally independent sensors.	Compare values of observables with predetermined or real-time calculated thresholds, number of desired object-like features matched, quality of feature match, mean and standard deviation of a measurable parameter, etc., for each object in the operational scenario. Analysis of the data offline determines the value of the likelihood function that expresses the probability that the data represent an object type a_j.
Dempster–Shafer evidential reasoning	Identification of events or objects $a_1, a_2, \ldots, a_n$ in the frame of discernment θ.	Identification of objects of interest, geological features, and other objects that can be detected by the sensors or information sources at hand.
	Probability masses m for individual events or objects, union of events, or negation of events reported by each independent information source (e.g., sensors, telecommunication devices).	$m_{S1} = \begin{bmatrix} m_{S1}(a_1 \cup a_2) = 0.6 \\ m_{S1}(\theta) = 0.4 \end{bmatrix}$ $m_{S2} = \begin{bmatrix} m_{S2}(a_1) = 0.1 \\ m_{S2}(a_2) = 0.7 \\ m_{S2}(\theta) = 0.2 \end{bmatrix}$
Artificial neural networks	Artificial neural network type	Fully connected multilayer feedforward neural network to support object classification.
	Numbers of hidden layers and weights	Two hidden layers, with the number of weights optimized to achieve the desired statistical pattern capacity for the anticipated training set size yet not unduly increase training time.
	Training data sets	Adequate to train the network to generalize responses to patterns not presented during training.

(continued)

Table 6.1 (*Continued*)

Data Fusion Algorithm	Required Data and Information	Example
Voting logic	Confidence levels that characterize sensor or other data source outputs used to form detection modes.	Sensor A output at high, medium, and low confidence levels (i.e., A_3, A_2, and A_1, respectively). Sensor B output at high, medium, and low confidence levels (that is, B_3, B_2, and B_1, respectively). Sensor C output at medium and low confidence levels (i.e., C_2 and C_1, respectively).
	Detection modes	Combinations of sensor confidence level outputs that are specified for declaring valid objects. Based on ability of sensor hardware and signal processing to distinguish between objects of interest and other objects or clutter.
	Boolean algebra expression for calculating sensor system detection and false-alarm probabilities	For a three-sensor, four-detection-mode system, System $P_d = P_d\{A_1\}\, P_d\{B_1\}\, P_d\{C_1\} + P_d\{A_2\}\, P_d\{C_2\} + P_d\{B_2\}\, P_d\{C_2\} + P_d\{A_3\}\, P_d\{B_3\} - P_d\{A_2\}\, P_d\{B_2\}\, P_d\{C_2\}$, where P_d is the detection probability. A similar equation is used to calculate the false-alarm probability of the sensor system by replacing the detection probability with the appropriate sensor's false-alarm probability at each confidence level.
	Confidence-level criteria that define characteristics of each confidence level	Confidence that sensors detect a *bona fide* object increases with, for example, the length of time that one or more features exceed some threshold, the magnitude of received signal, the number of features that match predefined object attributes, the degree of matching of the features to those of pre-identified objects, or having the measured speed of the object confined to predefined limits. Confidence that cellular transmissions or other communications are indicative of a valid object increases with the number of reports that identify the same object and location.
	Conditional probabilities that link the inherent object detection probability $P_d'\{A_n\}$ of Sensor A at the nth confidence level to the probability $P_d\{A_n\}$ that the sensor is reporting an object with confidence level n.	Compute using offline experiments and simulations; also incorporate knowledge and experience of sensor and fusion system designers and operations personnel.
	Logic-gate implementation of the Boolean algebra probability expression.	Combination of AND gates (one for each detection mode) and OR gate.
Fuzzy logic	Fuzzy sets	Object identification using fuzzy sets to specify the values for the input variables; e.g., five fuzzy sets may be needed to describe a particular input variable, namely, very small (VS), small (S), medium (M), big (B), and very big (VB). Input variables for these types of fuzzy sets include length, width, height, ratio of dimensions, speed, etc.

(*continued*)

Table 6.1 (*Continued*)

Data Fusion Algorithm	Required Data and Information	Example
Fuzzy logic (continued)	Membership functions	Triangular or trapezoidal shaped, although other contours are used. Shape parameters such as lengths of bases are determined through offline experiments designed to replicate known outputs for specific values of the input variables.
	Production rules	IF–THEN statements that describe all operating contingencies; heuristically developed by an expert based on experience in operating the object identification system or process.
	Defuzzification method	Fuzzy centroid computation using correlation–product inference.
Kalman filter	Object kinematic and measurement models	$\mathbf{X}_{k+1} = \mathbf{F}\mathbf{X}_k + \mathbf{J}\mathbf{u}_k + \mathbf{w}_k$, $\mathbf{Z}_{k+1} = \mathbf{H}x_{k+1} + \boldsymbol{\varepsilon}_{k+1}$, where **F** is the known $n_x \times n_x$ state transition matrix, **J** is the $n_x \times n_u$ input matrix that relates the known input driving or control function $\mathbf{u}_k$ at the previous time step to the state at the current time, **H** is the $n_z \times n_x$ observation matrix that relates the state $\mathbf{X}_k$ to the measurement $\mathbf{Z}_k$, and $\mathbf{w}_k$ and $\boldsymbol{\varepsilon}_k$ represent the process- and measurement-noise random variables, respectively.
	Process noise covariance matrix	If the tracked object is represented by a constant velocity kinematic model, the covariance matrix is given by $$\mathbf{Q} = q\begin{bmatrix} \frac{1}{3}(\Delta T)^3 & \frac{1}{2}(\Delta T)^2 \\ \frac{1}{2}(\Delta T)^2 & \Delta T \end{bmatrix},$$ where q is the variance of the process noise, and ΔT is the sampling interval.
Interacting multiple models	Object kinematic models, current probability of each model, and the model transition probabilities.	Model transition probabilities given by $$\mu_k^j \equiv \frac{\lambda_k^j \mu_{k=0}^j}{\sum_{l=1}^r \lambda_k^l \mu_{k=0}^l},$$ where r is the number of models, λ_k^j is the likelihood function of the measurements up to sample k under the assumption that model j is activated, and M^j is the event that model j is correct with prior probability $\mu_{k=0}^j$.

among confidence levels), and the ability to adequately model the state transition equation and measurement and process noise.

The two key concerns for data fusion are still:

- How does one represent knowledge within a computational database, particularly the information gained through data fusion?
- How can this knowledge or information be presented to a human operator in a way that supports the required decision processes?

> Data fusion is not the goal or end—the goal is to provide a human being the information necessary to support decisions such as traffic management options to divert traffic, inform travelers of delays, vary speed limits by lane, or change traffic signal timing in response to changes in real-time traffic flow.

References

1. M. P. Dana, *Introduction to Multi-Target, Multi-Sensor Data Fusion Techniques for Detection, Identification, and Tracking, Part II*, Johns Hopkins University, Organizational Effectiveness Institute Short Course ROO-407, Washington, D.C. (1999).
2. Notice of Proposed Rulemaking (NPRM), V2V Communications, 49 CFR Part 571 [Docket No. NHTSA-2016-0126], RIN 2127-AL55, Federal Motor Vehicle Safety Standards, National Highway Traffic Safety Administration, U.S. Department of Transportation, Washington, D.C., Dec. 13, 2016. https://www.federalregister.gov/documents/2017/01/12/2016-31059/federal-motor-vehicle-safety-standards-v2v-communications. Accessed Dec. 9, 2018.
3. "Harman and Samsung reveal 5G-ready connectivity," Autonomous Vehicle Technology Expo 2019 website. https://autonomousvehicletechnologyexpo.com/en/weekly-news.php?n=89327. Accessed Oct. 12, 2018.
4. J. R. Sayer, C. A. C. Flannagan, and A. J. Leslie, *The Cost in Fatalities, Injuries and Crashes Associated with Waiting to Deploy Vehicle-to-Vehicle Communication*, University of Michigan Transportation Research Institute, Mar. 1, 2018. http://umtri.umich.edu/sites/default/files/The%20Cost%20Associated%20with%20Waiting%20to%20Deploy%20DSRC.pdf. Accessed Dec. 9, 2018.
5. *C-ITS Platform Final Report*, January 2016. http://ec.europa.eu/transport/themes/its/doc/c-its-platform-final-report-january-2016.pdf. Accessed Aug. 26, 2016.
6. N.-E. El Faouzi and L. A. Klein, "Data Fusion in Intelligent Traffic and Transportation Engineering—Recent Advances and Challenges," Chapter

32 in *Multisensor Data Fusion: From Algorithm and Architecture Design to Applications*, H. Fourati, Ed., CRC Press, Boca Raton, Florida (2015).

7. L. Cholvy, "Information evaluation in fusion: a case study," *Proc. International Conference on Information Processing and Management of Uncertainty in Knowledge-Based Systems*, pp. 993–1000 (2004).
8. V. Nimier, "Information evaluation: a formalization of operational recommendations," *Proc. International Conference on Information Fusion*, pp. 1166–1171 (2009).
9. B. Khaleghi, A. Khamis, F. O. Karray, and S. N. Razavi, "Multisensor data fusion: A review of the state-of-the-art," *Information Fusion* **14**(1), 28–44 (2013).
10. *STANAG 2022: Intelligence Reports*, North Atlantic Treaty Organization (NATO), Brussels, Dec. 1992.
11. L. Cholvy, "Modelling information evaluation in fusion," *Proc. 10th International Conference on Information Fusion*, Quebec, 9–12 July, pp. 1–6 (2007).
12. A. E. Gelfand, C. Smith, M. Colony, and C. Bowman, "Performance evaluation of decentralized estimation systems with uncertain communications," *Proc. 12th International Conference on Information Fusion*, Seattle, 6–9 July, pp. 786–793 (2009).
13. M. Kumar, D. P. Garg, and R. A. Zachery, "A generalized approach for inconsistency detection in data fusion from multiple sensors," *Proc. 2006 American Control Conference*, Minneapolis, 14–16 June, pp. 2078–2083 (2006).
14. N.-E. El Faouzi, L. A. Klein, and O. De Mouzon, "Improving travel time estimates from inductive loop and toll collection data with Dempster–Shafer data fusion," *Transportation Research Record* **2129**, 73–80, Transportation Research Board, Washington, D.C. (2009).
15. D. S. Kim, J. D. Porter, M. E. Magaña, S.-J. Park, and A. Saeedi, *Wireless Data Collection System for Travel Time Estimation and Traffic Performance Evaluation: Final Report*, "Chapter 2-Literature Review," FHWA-OR-RD-12-13, SPR 737, OTREC-RR012-06, Oregon State University, Corvallis, May 2012. http://www.oregon.gov/odot/td/tp_res/docs/reports/2012/spr737_wireless.pdf. Accessed Jan. 20, 2016.
16. E. Blasch, R. Breton, and P. Valin, "Information fusion measures of effectiveness (MOE) for decision support," *Proc. SPIE* **8050**, 805011 (2011) [doi: 10.1117/12.883988].
17. L. A. Klein, *ITS Sensors and Architectures for Traffic Management and Connected Vehicles*, CRC Press, Boca Raton, Florida (2018).
18. J. Harding, G. Powell, R. Yoon, J. Fikentscher, C. Doyle, D. Sade, M. Lukuc, J. Simons, and J. Wang, *Vehicle-to-Vehicle Communications: Readiness of V2V Technology for Application*, "Chapter IX. V2V Communications Security," DOT HS 812 014, National Highway Traffic

Safety Administration, U.S. Department of Transportation, Washington, D.C., Aug. 2014. http://www.nhtsa.gov/staticfiles/rulemaking/pdf/V2V/Readiness-of-V2V-Technology-for-Application-812014.pdf. Accessed Dec. 9, 2018.

19. *Information Technology– Open Systems Interconnection– The Directory: Public-key and attribute certificate frameworks*, International Standard ISO/IEC 9594-8, Recommendation ITU-T X.509, International Telecommunication Union, Oct. 2012. http://www.itu.int/rec/T-REC-X.509-201210-I/en. Accessed Oct. 3, 2018.
20. V. Dragos and S. Gatepaille, "On-the-fly integration of soft and sensor data for enhanced situation assessment," *Procedia Computer Science* **112**, 1263–1272 (2017).
21. E. Waltz and J. Llinas, *Multisensor Data Fusion*, Artech House, Norwood, Massachusetts (1990).
22. E. Blasch, P. Valin, and E. Bosse, "Measures of effectiveness for high-level fusion," *Proc. 13th Conference on Information Fusion*, 26–29 July, Edinburgh (2010).
23. R. Falcon, R. Abielmona, S. Billings, A. Plachkov, and H. Abbass, "Risk management with hard-soft data fusion in maritime domain awareness," *Seventh IEEE Computational Intelligence for Security and Defense Applications*, 14–17 Dec., Hanoi, Vietnam (2014).
24. Y. Zheng, E. Blasch, and Z. Liu, *Multispectral Image Fusion and Colorization*, SPIE Press, Bellingham, Washington (2018) [doi: 10.1117/3.2316455].
25. D. L. Hall and A. Steinberg, *Dirty Secrets in Multisensor Data Fusion*, The Pennsylvania State University Applied Research Laboratory, State College, Jan. 2001. https://apps.dtic.mil/dtic/tr/fulltext/u2/a392879.pdf. Accessed Sept. 18, 2018.
26. D. McDaniel and G. Schaefer, "Real-time DBMS for data fusion," *Proc. Sixth International Conference on Information Fusion*, Cairns, Queensland, 8–11 July (2003).
27. D. Gettman, A. Toppen, K. Hales, A. Voss, S. Engel, and D. El Azhari, "Integrating Emerging Data Sources into Operational Practice—Opportunities for Integration of Emerging Data for Traffic Management and TMCs," FHWA-JPO-18-625, Kimley Horn and Associates, Phoenix, Arizona, Nov (2017).
28. D. Gettman, K. Hales, A. Voss, A. Toppen, and B. Tumati, "Integrating Emerging Data Sources into Operational Practice: State of the Practice Review," FHWA-JPO-16-424, Kimley Horn and Associates, Phoenix, Arizona, Nov (2016).
29. *Advancing Highway Traffic Monitoring Through Strategic Research*, Transportation Research Circular E-C227, L. A. Klein, Ed., Transportation Research Board, Washington, D.C., Dec. 2017.

30. L. A. Klein, *Sensor and Data Fusion: A Tool for Information Assessment and Decision Making*, Second Edition, SPIE Press, Bellingham, Washington (2012) [doi: 10.1117/3.928035].
31. N.-E. El Faouzi, H. Leung, and A. Kurian, "Data fusion in intelligent transportation systems: Progress and challenges—A survey," *Information Fusion* **12**(1), 4–10 (2011).

Appendix
The Fundamental Matrix of a Fixed Continuous-Time System

Following a derivation from R. J. Schwarz and B. Friedland, *Linear Systems*, McGraw-Hill, New York (1965), the differential equations governing the behavior of a fixed continuous-time system in vector-matrix form are

$$\dot{\mathbf{q}}(t) = \mathbf{A}\mathbf{q}(t) + \mathbf{B}\mathbf{x}(t), \tag{A.1}$$

$$\mathbf{y}(t) = \mathbf{C}\mathbf{q}(t) + \mathbf{D}\mathbf{x}(t), \tag{A.2}$$

where $\mathbf{q}$ is the state vector, $\mathbf{x}$ is the input vector or forcing function, $\mathbf{y}$ is the output vector, and $\mathbf{A}$, $\mathbf{B}$, $\mathbf{C}$, and $\mathbf{D}$ are constant matrices. Matrix $\mathbf{A}$ is a $k \times k$ square matrix, where k is the order of the system, $\mathbf{B}$ is a $k \times n$ matrix, $\mathbf{C}$ is an $m \times k$ matrix, and $\mathbf{D}$ is an $m \times n$ matrix.

The unforced (homogeneous) form of Eq. (A.1) is

$$\dot{\mathbf{q}}(t) = \mathbf{A}\mathbf{q}(t). \tag{A.3}$$

The solution to this system of equations will be shown to be

$$\mathbf{q}(t) = e^{\mathbf{A}(t-t_0)}\mathbf{q}(t_0) = \mathbf{\Phi}(t - t_0)\mathbf{q}(t_0), \tag{A.4}$$

where $\mathbf{q}(t_0)$ denotes the value of $\mathbf{q}(t)$ at $t = t_0$, and $\mathbf{\Phi}(t) = e^{\mathbf{A}t}$ is a matrix defined by the series

$$e^{\mathbf{A}t} = \mathbf{I} + \mathbf{A}t + \mathbf{A}^2\frac{t^2}{2!} + \mathbf{A}^3\frac{t^3}{3!} + \ldots \tag{A.5}$$

and is referred to as the *fundamental matrix* of the system. In engineering literature, $\mathbf{\Phi}(t - t_0)$ is called the *transition matrix* because it determines the transition from $\mathbf{q}(t_0)$ to $\mathbf{q}(t)$.

The series in Eq. (A.5) converges for all finite t and any $\mathbf{A}$. To demonstrate that Eq. (A.4) satisfies the differential equation [Eq. (A.3)], evaluate the time derivative of $e^{\mathbf{A}(t-t_0)}\mathbf{q}(t_0)$. According to Eq. (A.5), this time derivative equals

$$\begin{aligned}\frac{d}{dt}e^{\mathbf{A}(t-t_0)}\mathbf{q}(t_0) &= \frac{d}{d(t-t_0)}e^{\mathbf{A}(t-t_0)}\mathbf{q}(t_0)\\ &= [\mathbf{A}+\mathbf{A}^2(t-t_0)+\mathbf{A}^3\frac{(t-t_0)^2}{2!}+\ldots]\mathbf{q}(t_0)\\ &= \mathbf{A}[e^{\mathbf{A}(t-t_0)}\mathbf{q}(t_0)].\end{aligned} \tag{A.6}$$

Thus, Eq. (A.4) satisfies Eq. (A.3) subject to the given initial condition.

The initial value of the fundamental matrix is found as

$$\mathbf{\Phi}(0) = e^{\mathbf{A}0} = \mathbf{I}, \tag{A.7}$$

where $\mathbf{I}$ is the $k \times k$ identity matrix.

In Eq. (A.4), $\mathbf{\Phi}(t - t_0)$ is a matrix that operates on $\mathbf{q}(t_0)$ to give $\mathbf{q}(t)$. It is not necessary that $t > t_0$. The proof given above that $\mathbf{\Phi}(t - t_0)\,\mathbf{q}(t_0)$ satisfies Eq. (A.3) is also valid for $t < t_0$.

Thus, $\mathbf{\Phi}(t - t_0)$ permits calculation of the state vector at time instants before t_0, provided the system is governed by Eq. (A.3) during the entire interval defined by t and t_0.

The complete solution to Eq. (A.1) is obtained using the variation of parameter method. We assume that the solution is

$$\mathbf{q}(t) = e^{\mathbf{A}(t-t_0)}\mathbf{f}(t), \tag{A.8}$$

where $f(t)$ is to be determined. Then

$$\dot{\mathbf{q}}(\mathrm{t}) = \mathbf{A}e^{\mathbf{A}(t-t_0)}\mathbf{f}(t) + e^{\mathbf{A}(t-t_0)}\dot{\mathbf{f}}(t). \tag{A.9}$$

Substitution of Eq. (A.9) into Eq. (A.1) gives

$$e^{\mathbf{A}(t-t_0)}\dot{\mathbf{f}}(t) = \mathbf{B}\mathbf{x}(t). \tag{A.10}$$

Premultiplying by $e^{-\mathbf{A}(t-t_0)}$ gives

$$\dot{\mathbf{f}}(t) = e^{-\mathbf{A}(t-t_0)}\mathbf{B}\mathbf{x}(t). \tag{A.11}$$

Integration from $-\infty$ to t [assuming that $\mathbf{f}(-\infty) = \mathbf{0}$] results in

$$\mathbf{f}(t) = \int_{-\infty}^{t} e^{-\mathbf{A}(\lambda - t_0)}\mathbf{B}\mathbf{x}(\lambda)d\lambda, \tag{A.12}$$

allowing Eq. (A.8) to be written as

$$\begin{aligned}\mathbf{q}(t) &= e^{\mathbf{A}(t-t_0)}\int_{-\infty}^{t} e^{-\mathbf{A}(\lambda - t_0)}\mathbf{B}\mathbf{x}(\lambda)d\lambda \\ &= e^{\mathbf{A}(t-t_0)}\int_{-\infty}^{t_0} e^{-\mathbf{A}(\lambda - t_0)}\mathbf{B}\mathbf{x}(\lambda)d\lambda + \int_{t_0}^{t} e^{\mathbf{A}(t-\lambda)}\mathbf{B}\mathbf{x}(\lambda)d\lambda.\end{aligned} \tag{A.13}$$

The relation $e^{(u+v)\mathbf{A}} = e^{u\mathbf{A}}\,e^{v\mathbf{A}}$ is used in Eq. (A.13) and follows directly from Eq. (A.5).

Evaluating Eq. (A.13) for $t = t_0$ gives the initial state in terms of the input from $-\infty$ to t_0 as

$$\mathbf{q}(t_0) = \int_{-\infty}^{t_0} e^{-\mathbf{A}(\lambda - t_0)}\mathbf{B}\mathbf{x}(\lambda)d\lambda. \tag{A.14}$$

Thus, Eq. (A.13) becomes

$$\begin{aligned}\mathbf{q}(t) &= e^{\mathbf{A}(t-t_0)}\mathbf{q}(t_0) + \int_{t_0}^{t} e^{\mathbf{A}(t-\lambda)}\mathbf{B}\mathbf{x}(\lambda)d\lambda \\ &= \mathbf{\Phi}(t-t_0)\mathbf{q}(t_0) + \int_{t_0}^{t} \mathbf{\Phi}(t-\lambda)\mathbf{B}\mathbf{x}(\lambda)d\lambda.\end{aligned} \tag{A.15}$$

In fixed systems, it is usually convenient to set $t_0 = 0$. In this case, the fundamental matrix is $\mathbf{\Phi}(t)$.

Index

D

L

M

N

Lawrence A. Klein received the B.E.E. degree from the City College of New York in 1963, the M.S. degree in electrical engineering from the University of Rochester in 1966, and the Ph.D. degree in electrical engineering from New York University in 1973 along with the Founders Day Award for outstanding academic achievement.

Dr. Klein consults in sensor and data fusion applications to aerospace and traffic management; millimeter-wave radar system development for commercial, military, and homeland defense; and conceptual design of efficient multimodal ground transportation systems.

As Director of Advanced Technology Applications at WaveBand (now incorporated into Sierra Nevada Corp.) in Irvine, California, he led efforts to market and develop millimeter-wave sensors and data fusion techniques for aircraft, missile, and security applications. At Hughes Aircraft (now Raytheon Systems), Dr. Klein managed Intelligent Transportation System programs and developed data fusion concepts for tactical and reconnaissance systems. He worked as Chief Scientist of the smart munitions division at Aerojet (now incorporated into Northrup Grumman Corp.), where he designed satellite- and ground-based multiple-sensor systems. As program manager of smart munitions programs, he developed sensor fusion hardware and signal-processing algorithms for millimeter-wave and infrared radar and radiometer sensor systems. He also served as program manager of manufacturing methods and technology programs that lowered costs of GaAs monolithic microwave integrated circuit components. Before joining Aerojet, Dr. Klein was employed at Honeywell, where he designed polarimetric radar and passive millimeter-wave monopulse tracking systems. His prior work at NASA Langley Research Center explored remote sensing of the Earth's surface and weather from both phenomenology and sensor development aspects.

Dr. Klein regularly teaches courses in traffic management and data fusion for SPIE, UCLA, and the Harbin Institute of Technology in Harbin, China. He is a member of Eta Kappa Nu and has been a reviewer for several *IEEE Transactions* journals and foreign publications. As a member of the Highway Traffic Monitoring Committee and past member of the Freeway Operations Committee of the Transportation Research Board, he identifies research needs and concepts for applying data fusion to traffic management. He is a licensed professional engineer (New York). His other books are *Millimeter-Wave and Infrared Multisensor Design and Signal Processing* and *Sensor Technologies and Data Requirements for ITS* (both published by Artech House), *Sensor and Data Fusion: A Tool for Information Assessment and Decision Making*, Second Edition (SPIE Press), *ITS Sensors and Architectures for Traffic Management and Connected Vehicles* (Taylor and Francis), and the third edition of the *Traffic Detector Handbook* (Federal Highway Administration).